Partly Sunny

the weather junkie's
guide to outsmarting
the weather.

Alan Fields

WINDSOR PEAK PRESS

The Credits
Lead guitar, cowbell and inspiration by Denise Fields
Copy-editing and saxophone solo on "The Weather Channel" by
Lisa Fleck
Congas, vocals and interior design by Bob Schram, Bookends
Publication Design
Cover design and flugelhorn by Hugh Anderson of Archetype
Drums and back cover photograph of the author by Susan Fisher
Architecture and curved hardwood floors by Rob Fisher
Guacamole by Zolo Grill and Pete Pflum.

The cover photograph by Warren Faidley was taken in southeast
Arizona. The reddish sky is perhaps caused by ash fallout from Mt.
Pinatubo in the Philippines in June 1991. Photos in Chapters 3 and
4 are courtesy of The Weather Channel.

Guitar solo on "Slow Moving Cold Front" by John Hiatt
Bagpipes and percussion on "Blizzards" by Ric Ocasek
Finger cymbals and wah-wah sounds by Ben Fields

This book was written to the music of the Barenaked Ladies, which
probably explains a lot.

The original title of this book was "Buy Me, Please," but was
changed after consultation with Mark Ouimet, marketing guru for
Publisher's Group West.

Alan Fields appears courtesy of Howard & Patti Fields.

Special thanks to Connie Malco at The Weather Channel, and all
the weather buffs who contributed their insights to this book.

Extra special thanks to Denise Fields.

Distributed to the book trade and the stratosphere by Publisher's
Group West, 4065 Hollis St., Emeryville, CA 94608. 1-800-788-3123.
Thanks to the entire staff of PGW for their support.

To order this book, call 1-800-888-0385. Or send $12.95 plus $3
shipping to Windsor Peak Press, 1223 Peakview Circle, Boulder, CO
80302. Questions or comments? Please call the author at (303) 442-
8792. Or fax him a note at (303) 442-3744. Or write to him at the
above address in Boulder, CO.

Library Cataloging in Publication Data

Fields, Alan 1965-
    Partly sunny: the weather junkie's guide to outsmarting the
weather / by Alan Fields
    1st Edition
    256 pages
    Includes index.
    ISBN 0-9626556-5-1
    1. Weather 2. Meteorology. I. Fields, Alan. II. Title
    QC981.W24 1995.
    551.5—dc20

This book is dedicated to the memory of Harold Taft,
the meteorologist who inspired my interest in the weather.

*"He not only had weather enough, but weather to spare; weather
to hire out; weather to sell; to deposit; weather to invest; weather
to give to the poor."* —Mark Twain.

Overview

Introduction 1

Part One: Weird Science? Weather & the
Television 5

Weird Science looks at the strange mix of television and meteorology. First, there's a brief history lesson into how TV meteorology came to be. Then, we'll interview the Top 15 best local TV weathercasters and learn their secrets to outsmarting the weather. You'll discover why certain areas have quirky climates and how the pros forecast the weather. Next, it's a backstage tour of The Weather Channel, complete with biographies of the on-camera meteorologists and a look into the future of the all-weather network.

Part Two: Outsmarting the Weather for
Under $500 63

What's the most affordable weather station for your home? This section gives you the answer, as well as an in-depth look at weather instruments for business and educational purposes. Specific brands and companies are reviewed and rated. Next, we'll look at several other tools for outsmarting the weather, including weather radios, police scanners, and lightning detectors.

Part Three: Cruising the Weather
Information Superhighway 115

Learn how to download satellite pictures on your personal computer and more within this section. We start off with "easy" day-trips, including weather services available by fax and phone. Then, we'll cruise the weather databases on CompuServe and America On-Line and introduce you to weather software programs. Finally, you'll learn about earth satellite stations and military weather broadcasts. Each chapter includes a list of required equipment to do each task, as well as the degree of difficulty.

Part Four: The Best Gifts for the Weather Junkie 165

What do you buy a weather buff? How about a car thermometer, a special umbrella or a weather watch—this section gives you several ideas. Next, a complete wrap-up of the best mail-order catalogs for weather gadgets and gizmos gives you more gift ideas. A listing of organizations for weather buffs rounds out this section.

Contents

Introduction 1

◆ CHAPTER 1
Meet the Weather Junkie:
Why the Weather is an Obsession 1
A Serious Problem from an Early Age 1
Top 5 Reasons Why the Weather Is Addictive..................2
A Preview and Look Ahead 3

Part One:
Weird Science? Weather & the Television

◆ CHAPTER 2
The Top 15 Best Local TV Weathercasters...........................7
Weaned on TV Weather...7
The Ballad of Jack Van Roy 7
Weather: The Media Unplugged 8
Wooly Lamb and the Sordid History of Weather TV 9
A Clouded History: The Four Phases of TV Weather........9
How I Picked the Country's Best Weathercasters 10
The Top 15 Best Local TV Weathercasters 11
1. Vince Condella, WITI-TV, Milwaukee, Wisconsin 11
2. Dave Dahl, KSTP-TV, Minneapolis, Minnesota 13
3. David Finfrock, KXAS-TV, Dallas, Texas 15
4. Greg Fishel, WRAL-TV, Raleigh/Durham, NC 17
5. Doug Hill, WUSA-TV, Washington, D.C. 19
6. Janice Huff, KRON-TV, San Francisco, California.......20
7. Harvey Leonard, WHDH-TV, Boston, MA 22
8. Jim Little, KOIN-TV, Portland, Oregon......................23
9. Dave Murray, KTVI-TV, St. Louis, Missouri 25
10. Mike Nelson, KUSA-TV, Denver, Colorado.............27
11. Bryan Norcross, WTVJ-TV, Miami, Florida 29
12. Ed Phillips, KNXV-TV, Phoenix, Arizona..................30
13. Steve Pool, KOMO-TV, Seattle, Washington 32
14. Tom Skilling, WGN-TV, Chicago, Illinois.................34
15. Mike Thompson, WDAF-TV, Kansas City, MO.........36

◆ CHAPTER 3
Weather You Can Always Turn to:
Backstage at the Weather Channel 39
  Vegas is Not in the Eastern Time Zone 39
  Weather & Your Pork:
    The Early Years of Bac-Os and Losses 40
  The Hurricane that Saved the TV Network....................40
  The Weather War Room...41
  Para-Social Scores: Para-Normal or Para-Strange?
    Grading the Weathercasters: Map Blockage 42
    and the Big Cities..42
    Story Telling: The Weather as a Novel 43
  The 11:30 Briefing 43
  Waving at Air: Live in the Studio....................................44
  If I Programmed The Weather Channel 45
  The Local Forecast Soundtrack Album............................47
  Merchandising the Weather: Golf Balls & Turtlenecks ..48
  Forecasting the Future of The Weather Channel............48

◆ CHAPTER 4
Show & Tell: The Secret Lives of
The Weather Channel Meteorologists 53
  Bios & Bizarre But True Facts
    About the On-Camera Personalities 53

Part Two:
Outsmarting the Weather for Under $500

◆ CHAPTER 5
Weather Stations for Home, Business & School...................65
  No One Lives At the Airport 65
  The Three Basic Categories...66
  Weather 101: The Basic Function of a Weather Station..67
    Shopper Tips & Accuracy Advice for Five Instruments...67
  How to Order a Station 71
  Weather On The Homefront: Digital Stations
    for Fun & Education 72
    The Best: Davis Weather Monitor II 72
    1st Runner-Up: WeatherMax..72
    2nd Runner-Up: The Weather Report.......................81
    3rd Runner-Up: Fourth Dimension History Logging
      Station..83
    4th Runner-Up: Rainwise..86
  The 9-to-5 Weather Station: Industrial & Professional
    Applications 89
    The Best: MesoTech Automatic Weather Station 89
    1st Runner-Up: Capricorn...91
    2nd Runner-Up: Nimbus Weather Instruments.......92
  The Three R's: Rainfall, Relative Humidity and Record
  Temperatures...94
    The Best: Automated Weather Source 94

◆ CHAPTER 6
More Tools for Predicting the Weather......................................99
  Weather Radios: Bad Weather Turns Them On..................99
    What You'll Hear: Severe Weather Warnings &
      Three Other Broadcasts..101
    Well, Nobody's Perfect: The Shortcomings of the
      National Weather Service 103
    The Best Weather Radios 104
      Alert Models: Maxon, Midland, Radio Shack,
        WeatherOne 105
      Basic Models: Sony, Radio Shack......................106
  Scanners: Tuning In Your Town 107
    What You'll Hear: Weather Spotters & Six Other
      Frequencies 108
    Which Model Should You Buy?............................109
    Frequency Finders 110
  Lightning Detectors: Seeing the Storm Before It Sees You ...110
    Who Needs A Lightning Detector? 111
    How They Work..111
    The Ratings:
      Storm Alert, Stormwise Lightning Alert 111

Part Three:
Cruising the Weather Information
Superhighway

◆ CHAPTER 7
Easy & Fun Day Trips on the Weather Information
Superhighway..117
    Five Hundred Channels and the Weather's On............117
    📅 *Exit 1:* A Weather Buff's Guide to The Weather
      Channel..121
    📅 *Exit 2:* Weather By Phone—Talking Yellow Pages &
      Other Hotlines 121
    📅 *Exit 3:* Fax It: Zapping Weather Information to a
      Fax Machine..124

◆ CHAPTER 8
More Difficult Trails on the Weather Information
Superhighway..127
    📅 *Exit 4:* Personal Computers & Weather Databases....127
    📅 *Exit 5:* Brewing Up Home Forecasts.........................129
    📅 *Exit 6:* Weather On-Line—Using CompuServe,
      GEnie and America On-Line 130
    📅 *Exit 7:* Now What? Using Inexpensive "Shareware"
      Programs to Massage Data 133

◆ CHAPTER 9
Experts Only: The Most Challenging Journeys On the
Weather Information Superhighway.................................137
   What's Required:
      Souped-Up Computers and a Thick Wallet............137
   ☐ *Exit 8:* Weather Databases On-Line:
      WSI, Alden & Accu-Weather....................................138
      Other Weather Services that Won't Break the Bank .143
   ☐ *Exit 9:* Satellite Photos Without the Dish
      Tapping Into Free Military Weather Broadcasts.....145
   ☐ *Exit 10:* Talking to the Satellites: Earth Stations 146
      The Future: Doppler Radar at 55 mph and Three
      Other Soon-to-be Technologies..............................147

◆ CHAPTER 10
In the Kitchen With the Earth: Recipes for Tasty Weather
Phenomena..151
   How to Outsmart Various Weather Events...................151
      Thunderstorms:
         Popcorn Varieties & Squall Lines......................151
      Tornadoes:
         A Different Twist on the Same Old Dish..........153
      Heatwaves: Just a Lot of Hot Air? 156
      Blizzards: As Chilly as a Tasty Freeze......................158
      Hurricanes: The Big Blow.......................................161

Part IV:
The Best Gifts for the Weather Junkie

◆ CHAPTER 11
The Top 10 Gifts & More For Weather Junkies.................167
   Umbrellas to Weather Watches, Tornado Books to Auto
      Thermometers—What to Buy the Weather Buff....167
   The Best Weather Catalogs...171
   Organizations for Weather Buffs.................................175

Meet the Weather Junkie: Why the Weather is an Obsession

I LOVE THE WEATHER. Well, to be honest, I'm actually obsessed by the weather. In writing the introduction to this book, I had planned to tell you how I've been a weather junkie since age five. And how I grew up reading every book about the weather I could get my hands on. And how I dreamed of having a color radar set-up in my bedroom.

But, see, then there was this storm.

Right when I was supposed to compose this part of the book, a large storm blew up right over my house. Now, I should explain that as I write this, our house is being remodeled, and the upstairs is a sea of two-by-fours and exposed pipes. So, I'm standing there in the dark (because there's no electricity), with a weather radio in one hand and a police scanner in another, watching the lightning flash all around me. The storm is pummeling the town right below our house. Hurricane-force wind gusts bend the trees, and thunder crashes outside. I'm missing my deadline. And I couldn't be happier.

This is a sign of a person with a serious problem.

Sure, I could blame my weather habit on having grown up in Texas, where the weather is almost as big as football. But that wouldn't be quite fair—grow up anywhere in North America and you can experience incredible weather. For this, we have to credit our continent's unique position as a flash-point between cold Arctic air and warm air from the Atlantic and Pacific Oceans, plus the Gulf of Mexico.

No, there has to be some serious genetic defect at work here, some mutant X chromosome that transforms ordinary human beings into weather buffs. What else could explain an affliction that strikes at an early age and seems to get worse over time? Certainly, the great advances in meteorological technology have exacerbated the addiction—24-hour weather on cable, Doppler radar with 256 colors, home weather stations, and more.

So, I figured it was time I faced up to my addiction—and

celebrate it. This book is for all weather junkies and for those who'd like to get addicted but never knew how. First, let's take a look at the top five reasons why the weather is so addictive.

The Top Five Reasons
Why the Weather is So Addictive

1. IT'S SOMETHING CONGRESS CAN'T SCREW UP. Man has tried to tame the weather for hundreds of years, from rain dances to cloud seedings. It hasn't worked. The weather defies any effort by man to control or even predict it—sure, we've come a long way toward figuring out whether you'll need an umbrella tomorrow, but good luck trying to guess the number of Atlantic hurricanes that will strike the U.S. this year. On a daily basis, weather thumbs its nose at the most powerful nation on Earth. You've got to admire its chutzpah.

2. WEATHER IS OUR LAST CONNECTION TO NATURE. Ever walk around New York City? Every square inch of that island has been paved, concreted, and encased in steel. Yet despite the effort to blast the last bit of nature to New Jersey, the weather remains. Towering clouds billow above the tallest skyscrapers, reminding 10 million souls who's really in charge.

Of course, it's that way in the suburbs, too. You can manicure your lawn to perfection, but that won't stop a hailstorm from destroying your petunias. Our efforts to heat and "air condition" every home, building, and car only seem to disconnect us from nature; step outside the cocoon and the weather will always be there to greet you, whether you like it or not.

3. IGNORE IT AT YOUR OWN PERIL. Every time a hurricane threatens the U.S., the media always finds the obligatory idiot who proudly announces he's "riding out the storm." Two days later, the cameras return to find his house pulverized to bits, while the harried owner relates the story of how he and his cat "got out just in time."

Even if you don't find the weather interesting, it's darned hard to ignore. The weather is unique in its ability to sneak up on you when you're not looking.

4. FORECASTS ALLOW US TO SEE INTO THE FUTURE. Controlling destiny is sexy—and weather forecasts allow us to do just that . . . sort of. Think about it—it's next to impossible to predict what will happen with the stock market tomorrow, but we have a pretty good idea of what the weather will be. Those "extended outlooks" are like a crystal ball, enabling us to juggle a schedule to avoid a blizzard or to change a golf date when a rainstorm threatens. Psychoanalyze any weather buff and you'll likely find a modern day Merlin who delights in amazing friends and family with tales of what the future holds. Weather forecasts tease us with the possibility of controlling our own fate—we can see a few days ahead and are tantalized with the prospect of some day seeing further.

5. THE WEATHER ALWAYS CHANGES—BUT NEVER IN QUITE THE SAME WAY. "If you don't like the weather now, just wait five minutes, and it'll change." You'll hear that saying just about anywhere you travel in North America, and it's true up to a point. The weather *does* change, sometimes suddenly and, on occasion, violently. What makes the weather so fascinating is that these changes occur in a seemingly endless stream of variations—one summer can be 10 degrees hotter than normal while the next winter sees record snowfall. There's little rhyme or reason as to why a string of tornadoes pummels one area, leaving another area unscathed.

And that's what this book is all about—celebrating our fascination with the weather. First, I'll talk with the ultimate weather junkies: the country's best local TV forecasters, who've turned a passion for the weather into a full-time job. You'll learn how a local climate's quirks can throw a curve ball into forecasts and how the pros have learned to outsmart the weather.

Next, it's on to the Holy Grail of Weather Forecasting, a group of people who have forever transformed the business of broadcast weather. Yes, it's The Weather Channel. I'll take you on a backstage tour of the first all-weather network and look at what makes it tick.

After you learn how the pros do it, it's time to look at how *you* can outsmart the weather. I'll review and rate affordable weather stations, as well as explore other tools for outsmarting the weather.

Finally, it's off to cyberspace for a trip on the Weather Information Superhighway. You'll learn how to pull down weather information on a personal computer, zap maps to a fax machine, and tap into powerful weather databases. A "cookbook" on various weather phenomena in Chapter 10 shows how to pull it all together, using all of the above tools and tricks to outsmart the weather.

If that weren't enough, you'll even find my list of the best gifts for weather junkies, from a replica of the thermometer Galileo used to a watch that's a mini-weather station.

So, let's get on with it. Before another storm comes along to distract me.

Part One

weird science?
weather and the
television

2

The Top 15 Best
Local TV Weathercasters

I WAS WEANED ON TV WEATHER. It was at about the age of six that I realized each local station's weather came on at a slightly different time. With some fast remote control work, I discovered I could actually watch snippets from all three network affiliates—the current map from one channel, the radar from another, and finally the forecast from the third. That was three doses of weather in one half hour at 5 PM. And 6 PM. And 10 PM.

Sure, I drove my parents crazy, but at least they were never caught off guard by a storm.

There was something magical about each weathercast, even when the weather was boring. Maybe it was the maps—before the age of computer graphics, weathercasters actually had to be artists, drawing the highs, lows, and fronts with colorful chalk. A splash of green highlighted rain showers; a bright blue line with those little triangles sketched an advancing cold front. Today, this element of the weathercast is even more enticing with all the computer graphics and radar maps available. But even if you strip away all the fancy graphics and do-dad gadgets of today's weathercasts, you can't escape the one element that makes it the most popular part of any news show: the weathercaster.

I have always been fascinated with TV meteorologists. What a cool job. Getting to live, eat and breathe the weather for 40 hours a week AND pull down a paycheck. Next to being the official taster at Ben & Jerry's, I can't imagine a better career.

A pivotal event in my life was meeting my first weathercaster, Jack Van Roy. Jack worked at the ABC affiliate in Dallas, Texas, in the mid 70s, and I watched him religiously. When he came to speak to my second-grade class, I was in awe. A few years later, I ran across Jack again. This time it was in Sears—he was selling vacuum cleaners. He explained to me how the station unceremoniously dumped him when they found another hot-shot weathercaster. This hammered home an important rule about television weathercasting—it was the word "television" that always came first.

Maybe it was that encounter with Jack Van Roy in the vacuum department of Sears that convinced me not to become a TV weathercaster. I had visions of spending years in school, studying obscure topics like atmospheric physics, only to find myself ultimately working the drive-through window at Wendy's at age 53. TV probably has the world's worst retirement plan, second only to the Mafia's.

Despite this dose of media reality, I am still fascinated by TV weathercasters. I'm particularly amazed at the ones that become institutions—you know, the ones who will only leave the station when they retire or die. What is it that transforms an ordinary broadcaster into a local weather god (or goddess)?

That's what this chapter is all about. What makes a truly great weathercaster tick? Were they weather junkies from a young age, or did they get the bug later in life? One trait common to all of the best weathercasters is a real passion for the weather. And I mean *passion*. Get them talking for a minute about the local weather and you can't shut them up—they reel off the dates of snowstorms as though they were their children's birthdays and recite records of extreme temperatures like a grocery list.

This passion is missing from the rest of the news program. Watch any local newscast and you get the feeling that many of the other on-air "personalities" would rather be doing something else, but most weathercasters really seem to love reporting the weather. The same cannot be said for, say, the traffic guy.

Another common trait: most weathercasters are darned friendly people. Call the "weather center" at your local TV station, and you'll likely find that the chief meteorologist actually answers the phone. Try doing that with the sports anchor.

Maybe it's this accessibility that separates the beloved weathercaster from other media types—and perhaps creates the cult of personality that makes it difficult for a station to get rid of them before they die. Newspapers are full of stories of headstrong program directors who blow into town like an ill wind, determined to make their "mark" on the station and do so by announcing the firing of a popular weathercaster. After a public outcry like the kind reserved in previous years for unpopular wars, the program director backs down and is banished to the network's affiliate in Nome, Alaska. The weathercaster continues on, unscathed.

Except, that is, for Jack Van Roy. Sorry, Jack . . . somehow you came along too soon, before sainthood was bestowed on local weathercasters.

Weather, The Media Unplugged

Although most local newscasts are preprogrammed, scripted down to the last word of happy talk that ends the show, the weather segment is usually five minutes of ad-lib-bing and off-the-cuff commentary. It's like "News, Unplugged." Hence, the best weathercasters are able to stand there for five minutes and MAKE THIS STUFF UP AS THEY GO ALONG. This is obviously a rare skill. Take the teleprompter away from the average news anchor, and it can get ugly.

How most weathercasters are able to string together coherent sentences while simultaneously waving at the nonexistent image of a weather map is beyond me. Watch any rookie weathercaster doing his or her first broadcast on the weekend, and you'll probably agree that the six-figure salaries paid to the big guys are probably worth every last penny.

Wooly Lamb and the Sordid History of Weather TV

Television weather hasn't always been the province of polished meteorologists and souped-up Doppler radar. You can get a taste of the sordid evolution of TV weather by considering the following fact: The first television weathercaster was a lamb.

No kidding. A cartoon by the name of "Wooly Lamb" kicked off the century's first TV weathercast on October 14, 1941, on WNBT (later WNBC) in New York City. Each day, Wooly would sing, "It's hot. It's cold. It's rain. It's fair. It's all mixed up together. But I, as Wooly Lamb, will predict tomorrow's weather." A slide giving the forecast was then projected on the screen. From these sheepish beginnings, the business of television weathercasting was born.

A great book on this topic, *Television Weathercasting* by Robert Henson (McFarland & Company Publishers, $35; call 910-246-4460 to order), traces the history of TV weather from Wooly Lamb through The Weather Channel. After reading this, it becomes apparent that television weathercasting can probably be separated into four eras:

1. The Somber Stage. Before the 1950s, television was still in its infancy, and weather forecasts were usually read by serious newscasters. With the exception of Wooly Lamb (who continued forecasting for an amazing seven years), the weather segment of the local news was not awe-inspiring. Typical was Oklahoma City's WKY, which kicked off its first weathercast with an Air Force sergeant regurgitating the daily military weather briefing. As the novelty of TV wore off, however, this somber presentation would soon be history.

2. The Clown & Weather Girl Phase. Jokes and gags were the province of the TV weather in the 1950s, as television stations tripped over one another to "lighten up" the local news. Besides the clowns and puppets, another genre soon took to the airwaves: the "weather girl." Women in tight

9

sweaters were trotted out to boost the station's ratings among male viewers. This brief period at least opened the broadcast world's doors to women—several now famous media personalities got their start weathercasting, including Diane Sawyer, who started her media career as a "weather girl" in Louisville, Kentucky.

3. *The Age of Science and Tweed Jackets.* By the 1960s, the American Meteorological Society had launched a campaign to take back the weathercast from the pranksters. They created the AMS seal of approval and only granted it to "serious" weathercasters who met the group's guidelines for "professionalism." This effort paid off, and TV stations began hiring meteorologists in droves.

Unfortunately, this didn't necessarily improve the presentation of the weather. The scientists in tweed jackets and tortoiseshell glasses tended to dwell on isobars and other arcane meteorological terms, unconcerned that most people just wanted to know whether it was going to rain tomorrow. But there was no going back. When a series of strong storms in the 60s and 70s (from hurricanes to tornadoes, blizzards to heat waves) swept the country, the serious presentation of the weather was here to stay.

4. *The Bells & Whistles Stage.* The computer probably revolutionized TV weather more than any other development—snazzy graphics and "color" radar have transformed dull weathercasts into 3-D extravaganzas. The first computer graphics in 1981 were crude precursors of what was to come later in the decade. Despite the amateurishness of the early computerized maps, they were instant hits, and stations across the country rushed to equip their weather centers with the latest whiz-bang technology.

As a result, some "breakthroughs" were more hype than help when it came to explaining the weather. I still remember seeing 3-D satellite pictures for the first time and thinking, "So what?" Weathercasters could now smother themselves in a vortex of color graphics, but were they any better at predicting the weather?

How I Picked the Country's Best Weathercasters

In order to answer that last question, I interviewed weather enthusiasts across the country, asking them to explain why their favorite weathercaster was better than the competition. I should point out this was not exactly a scientific survey—weather buffs can be somewhat idiosyncratic in their picks, choosing a dark horse candidate over a popular broadcaster for a number reasons. Still, who can better evaluate a weathercaster than a weather junkie? At least weather junkies can tell which weathercaster knows his or her stuff—and which is faking it.

One common denominator among the best TV meteorologists: no brain-dead forecasts. Any idiot can get on television and read the National Weather Service forecast verbatim. It takes skill and chutzpah, on the other hand, to create your own forecasts, some of which might diverge from the party line. The best weathercasters have in-depth knowledge of the local climate—they've seen this storm set-up or that winter pattern before and know better than to rely solely on the prognostications from the National Weather Service.

Is your favorite weathercaster missing from this chapter? I hope to expand this survey in a future edition. Please feel free to contact me at (303) 442-8792 if you want to nominate a truly great weathercaster whom I somehow overlooked.

In alphabetical order, here are my picks for the Top 15 Best Local TV Weathercasters in the country. The numbers are not intended to be a ranking.

1 Vince Condella
Milwaukee, Wisconsin
WITI-TV

Vince Condella's first dose of television weathercasting was an eye-opening experience.

"I walked into the job interview and presented my resume to the news director," he recalled. The station in Madison, Wisconsin, was advertising for a part-time weekend weathercaster.

Vince certainly had the qualifications for the job. A native of Lombard, Illinois, he was working on a PhD in meteorology at the University of Wisconsin, after earning a master's degree there in 1979 and an undergraduate degree in atmospheric science from Purdue University in 1977.

"The news director took one look at my resume and tossed it into the trash," he said. "Then she said, 'All we want to see is how you look on TV.'" He was led down the hall to the studio and taped a sample weathercast. "I was terrible," Vince said. Nevertheless, he got the job and did his first weathercast in 1980. "I guess they figured no one watched on the weekends."

Despite the disastrous audition, Vince admits he was bitten by the TV bug. He'd always wanted to teach the weather and discovered television gave him a wider audience than he'd ever find in a classroom. He dropped out of the PhD program in 1982 and landed a job down the road at WITI in Milwaukee. "I was lucky," he said of his new position as the noon and 5 PM meteorologist for the popular station.

Vince's ascent to the top of Milwaukee's weather hill was aided by the station's number-one meteorologist, strangely enough. By all accounts, the head weathercaster at WITI just wasn't very good. "He wasn't very happy here. He would mispronounce names of towns and refer to the shore of Lake Michigan as the 'coast,'" he recalled. (The correct term is the "lake shore.") A mutual parting of the ways gave Vince his shot at the top spot, and he's been there (and in Milwaukee) ever since.

"I just love this city," Vince said, noting that while many weathercasters are always looking to move to bigger markets, he's been happy right there in Milwaukee. "This is a town that's relatively easy to forecast for. The only fly in the ointment is Lake Michigan."

It's this huge expanse of water that backs right up to Milwaukee's downtown that throws a spin on the area's climate. In the summer, cool lake breezes can form a miniature cold front that will chill the lake front while the suburbs bake. "It's not unusual to see a 20 degree temperature spread in just 20 miles," he said.

How does this happen? Well, the water temperature of Lake Michigan stays cool most of the year. In the summer, the sun will warm inland areas much quicker than the lake—the result is an inland area of low pressure that sucks in the cooler air that hovers over the lake. This "lake breeze front" is similar in effect to ocean breezes that cool coastal areas. Sometimes, the boundary between the cool, moist lake air and the warmer, dry inland air becomes a firing line for thunderstorms—this creates the potential for flooding since weak upper winds let the storms sit and dump huge amounts of rainfall.

In the winter, the process reverses itself. When a strong Canadian cold front pours into the area, the lake temperature is warmer than the surrounding area. As a result, downtown Milwaukee might see a "balmy" low of 32°F while the suburbs hover around 12°F. Easterly or northeasterly winds coming off the lake can greatly enhance snowfall amounts—during one storm last winter, the downtown was buried under a foot of snow while the suburbs were high and dry. As you can imagine, this can cause havoc for commuters.

Vince is looking forward to using the new NEXRAD technology to tame the wild-card effect that Lake Michigan plays on Milwaukee's weather. The new radar system has a "clear air mode" that shows wind direction even when there are no storms in sight. With this tool, Vince will be able to see the edge of the lake breeze front exactly, improving his ability to predict summer thunderstorms and winter storms.

"The secret to outsmarting the weather here is watching the wind direction," he emphasized. And perhaps watching Vince Condella isn't a bad idea either.

2 Dave Dahl
Minneapolis, Minnesota
KSTP-TV

Minneapolis gets so many snowstorms, folks there actually have different names for them.

"Perhaps the most common is what we call the Panhandle Hooker," explains Dave Dahl, the Twin Cities' most popular weathercaster. This storm may form near Denver and then swing south to the panhandles of Texas and Oklahoma, where it taps gulf moisture and gathers strength. Like a giant sling shot, the jet stream turns the storm north, where it heads straight for Minneapolis. The result? Up to 12 inches of snow blankets the Twin Cities.

But that's just the beginning when it comes to winter storms that hit Minneapolis. "Then we have the Dixie Dart, a somewhat rare storm that travels directly up the Mississippi," Dave said. Only a few of these storms hit the Twin Cities, but when they do, Dixie Darts can drop 18 to 24 inches of wet snow on the metro area.

Even more rare is a "Manitoba Mauler," a name Dave gives to storms that drop directly south from Canada. These intense storms bring tremendous ground blizzards to Minnesota, with 50 to 60 mile per hour winds. While Manitoba Maulers don't bring much snow (the drier air from Canada may result in just a few inches of dry, powdery snow), the winds whip the new snow and already accumulated amounts on the ground into drifts as high as five to eight *feet*.

And no discussion of Minneapolis snowstorms would be complete without at least a mention of the "Alberta Clipper," a fast-moving low-pressure area that forms in the same province of Canada from which it gets its name. Clippers can drop six inches of snow on the Twin Cities before quickly getting out of town.

Yet for all Dahl's knowledge of the area's winter storms, it was a close encounter with a tornado that kindled his interest in the weather. A native of Circle Pines, Minnesota (just 15 miles north of Minneapolis), Dave was ten years old the day in May 1965 when six tornadoes ripped though the Twin Cities, killing 16 people, injuring 750 more, and damaging 1000 homes and businesses. "I remember huddling with my five brothers and sisters in the hallway of my parents' home," he recalls. "We were there for two and a half hours." As the youngest child, Dave was at the bottom of the pile.

When his family emerged from their house, they realized how lucky they were. One of the strongest twisters (actually a twin tornado) came within a mere quarter mile of them. This funnel left a 20-mile path of destruction and was on the ground more than one hour.

"Obviously, this storm left a big impression on me," Dave said, recalling that after the tornado, he tried to read every book about the weather he could get his hands on. His parents encouraged him, buying him weather gadgets and experiment kits.

Dave thought he'd like to work for the National Weather Service after he graduated from Florida State University in 1977. Unfortunately, there were no job openings in the Minneapolis office at the time, and even worse, there were 45 qualified applicants in line in front of him. So he took a part-time job doing off-camera map preparation for the legendary Dr. Walt Lyons at KSTP-TV in Minneapolis. (A side note: he also worked part-time for Lyon's private weather forecasting company, Mesonet, where several of the country's best weathercasters also got their start.)

As fate would have it, one day in 1979 the weathercaster who did the morning news was sick, and Dave was tapped to do the broadcast. "I don't even own a suit," he protested to the news director to no avail. Consequently, Dave Dahl made his TV debut in a yellow short-sleeve leisure suit—with white shoes. "After that, they called me Banana Man for many years," he said. "I was so nervous, I couldn't even say my name. I had dry mouth. My lips stuck together. All in all, it was a disaster." Nonetheless, he somehow impressed the station's programmers and was chosen to fill in for the regular weathercaster from then on.

Dave continued to climb the weather ladder at KSTP during the 1980s, first taking over morning weathercasts, then weekends. In 1986, he became the head meteorologist and has been there ever since.

"I realize it's a cliche to say that you live in a town where the weather can change by the minute, but in Minneapolis, it really is true," he said. The confluence of three rivers (the Minnesota, Mississippi, and St. Croix) puts an unusual spin on the area's climate. "The river valleys can see significant snowfall amounts in the winter," Dave said. "In the summer, storms seem to follow the Mississippi and drift toward the Southeast."

The river valleys, which are lower in elevation than the surrounding terrain, seem to "stretch" the thunderstorms vertically, Dave explains. The result: the storms seem to energize, becoming more intense as they move toward Minneapolis.

It's this local expertise that makes Dave the number one pick as the best weathercaster by weather buffs in the Twin Cities. "I like the historical perspective he gives on different weather patterns," says one weather buff I interviewed. "He knows the people and city."

3 David Finfrock
Dallas/Ft. Worth, Texas
KXAS Channel 5

If there were one cardinal rule of TV weathercasting, it would probably be "never follow a legend."

Well, David Finfrock is my pick as the best TV weathercaster in Dallas/Ft. Worth, despite the fact that he is obviously breaking that rule. As the chief meteorologist for the NBC affiliate, Finfrock has a tough act to follow, replacing the legendary Harold Taft who died in September 1991 of stomach cancer.

Now, I grew up in Dallas, and there was probably no greater weathercaster than Harold Taft, who was undoubtedly present at the creation of weather. Taft did his first TV weather forecast in 1949 and worked for the same station (KXAS) for an amazing 40 plus years. No one was more beloved than Harold, whose folksy style and incredible weather knowledge made him a legend in a city that takes its weather very seriously.

In 1976, Taft hired the young Finfrock as a part-time weathercaster. Finfrock had just graduated the year before from Texas A&M's School of Meteorology. He remembers his job interview at KXAS quite well.

"I had never planned to get into television. This is all by accident," Finfrock said, explaining that he did research in Alaska after graduating from school and fully expected to get a job studying the polar ice caps in Antarctica. With an eye toward a career in research, Finfrock landed a fellowship at Texas A&M and planned to work on a master's in meteorology. "Three months into the fall semester, I got a call from Harold Taft."

Taft had plucked his resume from the National Weather Service's office in Ft. Worth and was impressed by the young meteorologist's credentials. "At the time, I had never heard of Taft, having grown up in Houston," Finfrock recalls, who went to Ft. Worth expecting an extensive interview process. Taft told him all he wanted to see was Finfrock giving the weather on camera.

"So they took me into the studio and sat me down in front of the maps left over from the noontime broadcast," he said. After giving a brief introduction, Finfrock was supposed to move to a map of Texas and give an impromptu talk on the day's weather. As he moved from the desk to the map, however, the chair he was sitting in went careening off the anchor platform, landing on the floor with a large crash. Unflustered, Finfrock went on to do his weather spiel, fully convinced he'd just blown the chance at being on TV.

One month later, however, Taft called to offer him the job. Years later, he explained it was Finfrock's cool reaction to the chair debacle that cinched the job. After fifteen years of doing the weather on the early morning, noon, and weekend newscasts, Finfrock took over for Taft when he fell ill in 1991. Since then, he's carved his own niche in the market.

What I like most about Finfrock is his continuation of Taft's strong emphasis on local weather knowledge—the weather in North Texas has more twists and turns than a roller coaster at Six Flags. Finfrock has been around long enough to develop a sixth sense about the Metroplex's weather . . . he's seen that cold front or this ice storm pattern before, and it shows.

One tricky weather scenario: predicting whether a slow-moving cold front will pass through the Dallas/Ft. Worth area. The key? Finfrock suggests looking at the barometric pressure in Salt Lake City, Utah, of all places. A cold front usually won't pass through the Metroplex until the barometric pressure is greater in Salt Lake City than it is in Brownsville, Texas (at the southmost tip of the state).

While the summer weather in Texas only comes in three flavors (hot, damn hot, and Son of Hot), winter forecasting keeps Finfrock on his toes. It doesn't snow much in North Texas, but when it does, it sends the populace into a major panic. As a result, there's incredible pressure on weathercasters like Finfrock to get it right. Complicating the prediction of snow is the particular geography of Dallas/Ft. Worth—it seems the area is always hovering around the freezing point in the winter, making a rain/snow prediction perilous at best.

Finfrock takes a page from Harold Taft's book when trying to outsmart a Texas snowstorm. "Harold once said that if you don't forecast snow until you see it falling, you'll only be wrong once." This beats the alternative of crying wolf and forecasting a blizzard several times, only to be wrong more often than not. Just like politics and skirt lengths, it's better to be conservative when you're forecasting snow in Texas.

"Sure, there are times when it looks like a clear-cut situation for snow in Dallas/Ft. Worth, and you just go for it," Finfrock said. "Unfortunately, I thought we had one of those clear-cut situations last winter—and it still didn't materialize."

4 Greg Fishel
Raleigh/Durham, North Carolina, WRAL-TV

Greg Fishel is in weather heaven—Raleigh/Durham, North Carolina. The area is home to several universities with good meteorology departments, including North Carolina State (which is right across the street from WRAL-TV). The university is on the cutting edge of meteorology research, and Greg has dabbled in several research projects to keep up to date on the latest climatological findings.

If that weren't enough, Raleigh/Durham is home to the state's headquarters of the National Weather Service, which recently moved its offices to the city. And even the Environmental Protection Agency has an office in town, staffed with dozens of meteorologists who are also doing weather research.

Greg has been in Raleigh/Durham for so long (since 1981), some folks there mistake him for a native, but he actually grew up in Lancaster, Pennsylvania. He got bit by the weather bug early in life. "As a kid, I remember being obsessed with snow," he says. "It always seemed like Lancaster got rain when the rest of the state was getting snow. Now looking back at this experience from living all this time in Raleigh, Pennsylvania seems like the Arctic Circle."

Unlike other weather buffs, however, Greg was not a fan of thunderstorms. "I was petrified of severe weather," he recalls. "As soon as I heard there was a tornado watch, I would freeze and watch the clock. As soon as the watch ended, I felt like a boulder was lifted off my shoulders."

Somehow, Greg channeled this fear into scholastic interest in the weather, and he graduated with a degree in meteorology from Penn State in 1979. His first job was at a private weather forecasting firm in Chicago, where one of his co-workers was John Coleman (later of The Weather Channel and "Good Morning America" fame).

Greg's big splash into television was at a brand new TV station in Salsbury, Maryland, in 1980. "Apparently, the general manager there was dumb enough to hire me," Fishel said, adding that his only TV experience to date was one weathercast on a small university student-run station. The Maryland station made a gallant attempt at staging a professional weathercast, including building a huge map of the

United States for Greg's broadcasts.

"There was only one problem," Greg recalls. "No one was watching." The station was up against an established competitor whose newscast had been on the air for years. "The biggest rating we ever got was a two share," he said—yes, that means only two percent of the folks in Salsbury were ever watching the station. Meanwhile, the other TV station had a 67 percent share of the audience.

Things were looking a bit grim in 1981 when Greg got a call from a WRAL-TV in Raleigh/Durham. They were looking for a weekend weathercaster who would add a little more science to their newscasts. Apparently, a reporter at WRAL who had worked with Greg in Maryland had recommended him to the news director. He gladly accepted the job and took the drive south in June 1981.

The head meteorologist at WRAL during those years was Bob Debardelaben, whose promos advertised him as the "biggest name in weather." When Bob finally retired in the summer of 1989, Greg took over as chief meteorologist.

"Raleigh/Durham's unique geographic location at the border of the Piedmont foothills and the coastal plain makes forecasting a challenge," Greg said. The problem? In the winter, a very shallow layer of cold air can get trapped in the Piedmont foothills. Meanwhile, the relatively warm water of the Atlantic ocean will feed mild southeast breezes across the area—the result can be a 20 degree difference in temperatures in just 20 miles. "Towns like Wilmington can get up into the 60s, while Raleigh shivers in the 40s, and there are 30s and frozen precipitation just to our west."

The area sees an average of seven inches of snow each winter but hasn't seen a "normal" year of snowfall since 1989. When asked if the absence of snow has lulled Raleigh into a false sense of security, Greg replies, "I hope so."

Perhaps the most memorable weather event during Greg's tenure in Raleigh/Durham is the massive tornado outbreak in March 1984, the worst in 100 years. A strong jet stream clashed with an unusually deep low pressure system that swept across the state—the result was a spate of tornadoes that killed 59 people in South and North Carolina. "Up until that time, I thought of severe weather as a big video game on the radar screen. Watching the death toll climb was a sobering experience."

5 Doug Hill
Washington, D.C.
WUSA-TV

Doug Hill is often wrong when he forecasts the temperature for Washington, D.C. And he's the first to admit it. "The biggest quirk to the weather here is the fact that the 'official' temperatures are taken at National Airport." This small airport is located on a strip of land in the middle of the Potomac River. In the spring, the water temperature is still in the 50s, and this cools the air at the airport. As a result, the "official" temperature is often 15 to 20 degrees cooler than the actual temperature at, say, 1600 Pennsylvania Avenue. Doug will forecast highs in the 80s, only to have to report that the official high at National was a chilly 68°.

Despite this disparity, weather buffs still give Doug high ratings. "He's the most straightforward weathercaster in D.C.," one told me, adding that "he doesn't get overly technical or alarmist. And, best of all, he also is more accurate forecasting winter blizzards."

And it's a good thing. "This city panics when it snows," Hill says, noting that a storm that dumps just three to six inches can shut down the nation's capital. Why? Doug speculates that perhaps it's the transient population of politicians from "everywhere but here," changing every four years with the shifting political winds.

Doug is a local boy, growing up in Towson, Maryland. While he attended college at Maryland State and the University of Maryland, he never earned a degree. "I don't have the most honorable collegiate record," he admits. Doug's lack of a college degree makes him somewhat of a dark horse choice as Washington's best weathercaster, especially given that he competes against popular meteorologist Bob Ryan at WRC. Yet even without the sheepskin, Doug has left his mark on the local weather scene. He started in Richmond, Virginia, and did the weather in Detroit at WXYZ before coming to Washington, D.C., in 1984. His twelve plus years at WUSA have given him a unique perspective on the region's weather.

"One of the things the computer models always miss is the effect of the Appalachian mountains just 30 miles to the west on our weather," he explains. Even though the mountains reach a height of just 4000 feet, this is high enough to create down-sloping winds in the spring and fall. The result:

high temperatures in the city can be as much as 10 degrees warmer than originally predicted.

Another tricky aspect to the weather in Washington, D.C., is the effect of the Chesapeake Bay and Atlantic Ocean. "Winds that channel up or down the bay can greatly affect temperatures and precipitation," Hill notes. In the summer, the bay has a channeling effect on thunderstorms, sending them careening over the same areas day after day, only to leave other areas high and dry.

From thunderstorms in the summer to ice storms in the winter, Washington, D.C., always seems to be in just the right position to get the worst of the weather. "The only thing we're missing is volcanic eruptions or plagues of locusts," Doug says with a laugh. "On second thought, maybe I shouldn't rule those out."

6 Janice Huff
San Francisco, CA
KRON-TV

Janice Huff's fascination with the weather started as a little girl growing up in Columbia, South Carolina.

"My grandfather was an avid outdoorsman," she explained, crediting his influence for her fascination with the weather. Each summer, Janice and her grandfather would sit outside on their porch and watch the thunderstorms roll in. During the hot, sticky South Carolina summers, it was the thunderstorms that provided the only relief—in a sense, a form of natural air-conditioning.

"Each summer, it seemed like the storms would form in a particular part of the sky," Janice said. Her grandfather called it "catfish corner," since the dark sky resembled the muddy water that catfish call home.

As the storms drew closer and the lightning and thunder became more intense, she would try to outlast her grandfather on the porch on those summer evenings. "He would always win," she said. "I'd always end up under a bed, as the thunder crashed outside. He'd always ride out each storm on the porch, no matter how bad the wind, rain, or lightning."

Watching a thunderstorm is a rare occurrence for Janice today. As the meteorologist for KRON-TV in San Francisco, she only sees three to five thunderstorms per year. "And when it does thunder and lightning, it leads the news here," she said, laughing. Such storms are so rare in San Francisco that children can grow up there without ever hearing a clap of thunder; hence, when the rare storm occurs, it sets everyone off in a major panic.

Janice started her weather career as a student trainee for the National Weather Service after high school. After earning a degree in meteorology from Florida State University in 1982, she expected to have a career working as a meteorologist for the government. However, a federal hiring freeze forced her to change career paths.

Her friends encouraged her to go into television, but she thought they were nuts. Nevertheless, she took one course in broadcasting and got hooked. Upon graduation from college, she had two job offers and took the first one, a weather slot at a station in Chattanooga, Tennessee. Her career took her to Columbus, Georgia, and St. Louis, Missouri, before she landed the job at KRON in San Francisco in 1990.

Weather buffs like Janice's weathercasts for several reasons. "First, she's very knowledgeable," one told me. "I especially like the pollen and pollution reports. She even gives ozone forecasts."

But how hard can it be to forecast weather for a city that will probably never see a blizzard or tornado?

"People say the weather here is boring, but if they lived here just for one week, they'd change their mind," she said. The reason? "The Bay Area has a 1001 microclimates," she said, noting that temperatures can vary 40 to 50 degrees from the coast to inland areas. Inland towns like Antioch or Concord can see wide temperature swings from 60°F in the morning to over 100°F in the afternoon; meanwhile, beachfront towns can be stuck in the 50s all day.

And then there's the fog. "Have you ever seen the fog roll into San Francisco? It's absolutely fascinating . . . like a dry ice science experiment," Janice said. Each summer, the fog rolls in off the ocean, bathing parts of the city in gloom, while other areas are warm and sunny. Trying to forecast the weather for a city where the temperatures can vary dramatically in a matter of blocks can be tricky.

"You can always spot the tourists in San Francisco," she said. "They're the ones in shorts and beach attire in August, shivering in the foggy, cool weather." Actually, "summer" in San Francisco is in September and October, right after the fog dissipates and before winter storms bring rain. "Whatever the weather is in other parts of the country, it's the opposite here in San Francisco," she said. "The same goes for politics."

7 Harvey Leonard
Boston, Massachusetts
WHDH-TV

During the summer, Harvey Leonard is by all accounts a normal person. He talks with his family, goes shopping, and may even take a vacation. But when the seasons change, something happens to Harvey.

"I'm obsessed," he says. "I love winter storms."

Harvey grew up in the Bronx, New York, absolutely fascinated with snow and winter weather. "I'd get out there and measure the temperature on the patio when it was snowing. I would even volunteer to take the trash out—anything to get outside and see the winter storm."

At this early age, Harvey began building what he calls his "mental encyclopedia of winter storms," remembering the details of how much snow fell and which way the storm blew in. This obsession continued through his college years, as he earned not one but two degrees in meteorology (a bachelor of science from City College of New York and then a master's of science from New York University).

Harvey couldn't have picked a better place to watch winter storms than Boston. After a job weathercasting in Providence, Rhode Island, he joined WHDH-TV in May 1977. He's been at the same station for 17 years.

"This is probably one of the trickiest places to forecast winter weather," he says, explaining that Boston's proximity to the Atlantic Ocean makes predicting snow difficult. Warm winds off the ocean can melt a winter storm into a harmless rain in the city of Boston, while the suburbs are getting blasted with heavy snow. The station's large viewing area (from New Hampshire to Cape Cod) means weathercasters are expected to get it right not just for Boston, but for the entire region.

Harvey likes to do post-game analysis on winter storms, trying to figure out why one storm zigged while another one zagged. As a result, he's got a good track-record in predicting the rain/snow line, which can shift with the slightest change in winds. It's this pinpoint accuracy that has earned him legions of fans in the Northeast.

How does Harvey do it? "Before I look at the computer forecast models, I like to look at the current conditions, satellite imagery, and other data. Given the current weather, I like to guess what the computer model will say." At times,

Harvey has predicted storms that the computer missed.

One storm during the famous winter of 1993-4 provides a good example. The computer models said Boston would see little snow, but Harvey noticed a disturbing pattern that reminded him of an earlier storm that dumped snow on the city. After consulting his "encyclopedia," he decided to forecast 8-15 inches of snow—a major dump even for a snow-hardy city like Boston. What happened? What do you think? Harvey was right again.

Of course, there's more to the weather in Boston than just the eleven months of winter. Hurricanes can visit the area, as Gloria did in September 1985 and later Bob did in August 1991. The potential for wind damage and coastal destruction keeps weathercasters on their toes all year round. But try to talk about other types of weather in Boston, and Harvey will steer the conversation back to his favorite topic: snow.

"The worst blizzard in Boston was February 6 and 7, 1978," he recalls. Boston got blasted with 27 inches of snow and tremendous coastal devastation. Friends asked Harvey (who was just newly hired at the station) whether this was a bad omen. "On the contrary," he said. "I thought this was a good omen. I love this stuff." And, of course, Harvey had nailed the forecast.

8 Jim Little
Portland, Oregon
KOIN-TV

Jim Little was not a weather buff growing up in Portland, Oregon. "I was a free-lance science enthusiast," he says, and was a big fan of the space program.

That changed on Columbus Day, 1962. That day one of the biggest windstorms to strike the Northwest in recent history roared into town and made quite an impression on a 12-year-old boy who would later become one of Portland's most popular weathercasters.

"I remember walking home from school that afternoon," Jim recalls. A light rain was falling, and he could tell a bigger storm was brewing. "We were watching cartoons when the big station here, KGW, went off the air. We heard on the radio that the station had its transmitter blown down, so my brother and I ran outside to see what was happening."

"As the wind was racing outside, I remember a piece of

sheet metal flying about 50 feet above our heads. Then my mother came outside and yelled, 'Get your butts in here!'"

The Columbus Day storm (with its gusts clocked to 160 mph) probably helped spark Jim's interest in the weather, although he majored in mathematics when he went to school on an Air Force ROTC scholarship at Linfield College in McMinnville, Oregon. After graduating, he spent five years on active duty in the Air Force, which sent him to the University of Texas in Austin to study meteorology.

His stint in the Air Force took him to bases in Tacoma, Washington, and Guam before landing at the Strategic Air Command in Omaha, Nebraska. There Jim got his first experience presenting the weather, doing morning weather briefings for four-star generals.

"You had to know everything," Jim said, adding that he'd study the maps for hours before doing the briefing. The generals loved to hassle him over the slightest missed forecast; one required him to give the forecast for each and every Air Force football game.

After the grilling from the generals, Little found television weathercasting to be a piece of cake. His first job was at a station in Twin Falls, Idaho, in 1976—an Air Force buddy was now the news director there and gave him his first shot in TV.

Jim learned all he could about Idaho weather before his first broadcast, expecting the anchor to quiz him about the weather like the generals did in his Air Force years. "Yet after my first weathercast, the anchor just turned to me and said, 'Thanks, Jim. And, after this commercial break, we'll have sports.' I remember thinking, 'Wow, this is easy.'"

Jim's weathercasting career has taken him to several markets, including Salt Lake City and Sacramento, before landing him back in his hometown, Portland, in 1979. Jim did the evening weather for KGW for 11 years before jumping to KATU in 1990 and finally to his current station (KOIN-TV, the CBS affiliate) in 1993.

Portland's unique position at the mouth of the Columbia River Gorge (which runs east/west) and the Cascade Mountains to the west makes for some fascinating (and tricky) weather. "Perhaps 90% of our weather comes from the Pacific, but there are systems that slip in the back door," he said.

In the winter, a piece of a large arctic high that is entrenched over the Rocky Mountains can slip into the Columbia River Gorge. This cold air follows the canyon and dumps out in Portland, chilling the eastern suburbs with temperatures in the mid 20s and sometimes frozen precipitation. Yet this shallow layer of cold air can't rise over the "west hills" of Portland, leaving towns like Beaverton, Oregon, with weather that is 15 to 20 degrees warmer.

"This happens once or twice a year," Jim explains. One quarter of an inch of snow will shut down the city, he said, except of course for the "banana belt" of Beaverton, which is always spared. "Every time this happens, I have to take dozens of calls from irate school children in Beaverton who want to know why school has been cancelled for every district except theirs."

When moist Pacific air flows over the arctic layer, freezing rain can form, sending Portland into a tizzy. This glaze ice is know as a "silver fog" in the Pacific Northwest.

"East winds are the wild card here," Little said. In the winter, east winds can bring cold air down the Columbia Gorge. Yet in the summer, east winds are warm, down-sloping breezes from the Cascades—this can send the mercury in Portland to over 100°F. "And that's rather miserable for a town that doesn't have much air-conditioning."

9 Dave Murray
St. Louis, Missouri
KTVI-TV

If you want to get into television weathercasting, you'll probably need an audition tape. For most of this chapter's top weathercasters, these tapes will hardly make it onto their career highlight reel—most are amateurish productions at a college TV studio or even worse.

That's not the case with St. Louis' top weathercaster, Dave Murray. His audition tape very well may make it to the weathercasters' hall of fame, if there ever is such a place in the future.

"A friend and I did a piece of film animation of a front moving across Wyoming," he recalls. He was working on a master's degree in meteorology at the University of Wyoming at the time. The process was similar to "claymation," where Murray moved the front, clouds, and map symbols, took a frame of film, and then moved the elements again. While the result probably didn't look as slick as today's computerized graphics, it apparently caught the eye of a news director at KSD-TV in St. Louis, who hired him in 1976.

Actually, Dave's audition tape garnered him two job offers: besides the St. Louis station, he was offered a spot on Boston's WBZ weather team. Ironically, that was one of the stations Dave had watched growing up in Providence, Rhode Island. "I used to drive my parents crazy," Dave said. "Since we could get both newscasts (from Providence and Boston), I would watch all six weather broadcasts each evening."

Dave's early interest in astronomy eventually gravitated into a passion for weather. In the sixth grade, he became a weather watcher for WBZ meteorologist Bob Ryan (now in Washington, D.C.).

Unlike other weathercasters, who got into television by accident, TV was always Dave's goal. "I'm not sure why, but it was the part of the weather that I always wanted to do." After receiving a degree in meteorology and geography from the University of Rhode Island, he went to the University of Wyoming and worked on urban climate research. It was that project that brought him to St. Louis.

"I took the job offer at KSD-TV in St. Louis since I had a better chance of becoming the head meteorologist there," Dave said. And he apparently was right. Just two weeks after joining the station, he was promoted into the lead spot.

But that was just the beginning of Dave Murray's fast-track weather career—in 1983, he got a call "out of the blue" from the executive producer of ABC's "Good Morning America." A job offer to work at the network followed, and he went off to New York City to take one of the most visible weathercaster positions in the country.

"It was quite a ride," Murray recalled. Sitting on the couch next to Joan Lunden and Charlie Gibson, he met just about every big celebrity, sports figure, and Hollywood star you could name. Yet after just three years, Murray called it quits.

"If I did that show for another three years, I'd be dead," he said. The grind of preparing the maps for three hours of weathercasts plus the one-hour commute from his home in Connecticut meant a 1 AM wake-up call every day. "I missed three years of my son growing up," he explained. "And besides, I wanted to get back to real weather." Dave's quick weathercasts on "Good Morning America" were seen by millions of viewers, but he missed tracking severe weather. "Come April and May, I would have thunderstorm withdrawal."

Murray has no regrets. "I still talk with Joan Lunden every now and then," he said, adding that "many folks still recognize me from that show." After leaving the network, he worked for WBZ-TV in Boston for three years—the same station he watched growing up and from which he spurned a job offer in 1976. Yet for all his time on the East Coast, he realized the place he really missed was St. Louis.

His opportunity to go back where he started came in 1989 when the general manager of ABC affiliate KTVI called to offer him a job. "I really wanted to return to St. Louis. It's a great place to bring up a family. My commute is just a few minutes so I can have dinner with my family every night. Besides, St. Louis is a four-seasons town, and lots of fun to forecast."

One of the quirks to the weather in St. Louis that Murray finds interesting are "disappearing" spring thunderstorms.

"A tremendous amount of thunderstorms will fire off to our west, move toward the city, only to die when they enter the metro area," he explains. Then the storms redevelop to the east of town and move on to terrorize Illinois.

Why does this happen? "St. Louis creates quite an urban heat-island effect," he said, noting the city is much warmer than surrounding areas. Rural counties to the west of St. Louis have much more vegetation, which contributes to their more humid and slightly cooler climate. When storms move toward the city, they encounter the dry air and usually piddle out. "The most severe storms we see are the ones that develop right on top of the city, while cold front-generated squall lines are usually disappointing."

Murray's experience doing the weather on both the East Coast and in the Midwest has given him an interesting insight into how different parts of the country view the weather—and a few funny stories as well. "Television producers in New York think the weather is a necessary evil . . . a two-minute space they'd rather use for something more important," he said. "The only exception is when a storm hits New York—then it's a big deal." During the first winter he worked for "Good Morning America," he remembers getting a call from a producer early one morning. "'We want you to go outside and do a live remote to show the major snow we got last night,' he said. I said, 'What major snowfall?'" The city had picked up a whopping two inches of snow the night before . . . and much of it was melting as the morning rush hour traffic began. "Nonetheless, there I was that morning, reporting on the major snow drifts that had struck New York City," he said with a laugh.

10 Mike Nelson
Denver, Colorado
KUSA-TV

Mike Nelson is definitely Mr. Smooth.

Looking like a weathercaster's answer to Dudley Do-Right with his square chin and curly locks, Nelson combines an impressive set of graphics and weather tools with a wide-eyed enthusiasm for the weather. The result is a polished presentation that is among the best in the country.

Nelson's latest toy is the "Skyscape Computer," which takes viewers on a trip "flying" around the country to check weather systems, temperatures, and clouds. Nelson looks right at home in the whirl of high-tech graphics—and it should be no surprise since that's how he got his start in the weather.

After graduating from the University of Wisconsin at Madison, Nelson got a job with ColorGraphics, a company that pioneered computer graphics for television weather-casts. In 1981, ColorGraphics unveiled the "Liveline1," a weather graphics system that ran off an Apple II computer. While the system was quite primitive by today's standards (a camera had to be trained on the computer to broadcast the crude graphics), it was a hit. And Mike Nelson was one of the company's top salesmen.

Together with his wife Cindy, the young meteorologist traveled the country, pitching the graphics system to televi-sion stations. One of their stops was KUSA-TV (then KBTV) in Denver, Colorado, where Nelson met famous meteorolo-gist Leon "Stormy" Rottman. "Stormy was very leery of the computer," Mike recalls.

While Stormy was a tough sell, it was Mike and Cindy Nelson who were easily sold—on Denver. "We said to our-selves that if the opportunity ever presented itself, this would be the ideal place to raise a family."

The opportunity came in 1991. After Stormy retired in 1987, KUSA-TV was looking for a strong newcomer to shake up its weather center. They found Mike Nelson, who at the time was probably the most popular weathercaster in St. Louis. Mike's six-year run at KMOX-TV in St. Louis was high-lighted by two Emmys and a slew of praise for his involve-ment in the community.

It is community service that separates Mike Nelson from other weathercasters—he spends countless hours each week giving "weather talks" to local elementary schools.

"When I was in school, it was a visit from a local weather-caster that sparked my interest in science," he said, hoping he can now inspire the next generation of scientists. Nelson has developed a whole skit to entertain the children, including "hail juggling" and the soon-to-be craze "the tornado dance."

Entertaining eight-year-olds is arguably easier than trying to forecast the weather for Denver, one of the toughest mar-kets in the country. Nelson said the Continental Divide, which soars to over 14,000 feet just west of the city, throws quite a curve into forecasting.

"Storms tend to break up over the mountains," he said, only to reform out on the eastern plains. Where a low pres-sure area forms greatly determines snowfall amounts in the winter, he pointed out. This quirk can leave an area like Colorado Springs (50 miles to the south of Denver) basking in warm, dry weather while Denver is getting socked with a foot of snow.

The tricky winter weather is just the beginning. "Denver

is one of the few places is the country where the weather can be life-threatening all year round." In the spring and summer, heavy thunderstorms can roll out of the foothills, dumping hail, heavy rain, and even a tornado on the plains. If that weren't enough, the occasional windstorm in January (caused by a high pressure in Utah and a low pressure in Kansas squeezing the air down the mountains) can create gusts of over 100 mph.

11 Bryan Norcross
Miami, Florida
WTVJ-TV

Bryan Norcross holds the distinction of being the only meteorologist in this chapter who's been the subject of a television "movie of the week." Millions of viewers saw Norcross (portrayed by an actor) become lionized as the savior of Miami during Hurricane Andrew.

Given his fame, few would guess that Norcross only became a television meteorologist after an early mid-life crisis prompted a career change.

His first job in TV was as an engineer, of all things. After growing up in Melbourne, Florida, he attended Florida State but didn't study in their famous meteorology program. He spent his undergraduate years majoring in mathematics and physics, and, for variety, was a disc jockey at a local radio station.

Norcross put his interest in science and the weather on hold while he drifted through a series of production and management jobs at local television stations. He eventually landed in Atlanta, where he worked at Turner Broadcasting Service (TBS) network as an executive producer.

"By the time I turned 30, I needed to do something different," Norcross said. So, he returned to Florida State, got a master's in meteorology and communication, and turned his passion for science into a second career as a television weathercaster. His career included brief stints at WXIA in Atlanta as well as doing weekend weather for a fledgling cable network by the name of CNN.

Yet it was Norcross' tenure in Miami that would take him to weather super-stardom. He started off at WPLG in 1983 as the weekend and then 5:30 weathercaster, where he created his trademark "neighborhood weather" segment: every day, he'd do a live remote broadcast from a different part of town.

"In Miami, very few people know anything about this town," he said. This "walk and talk" style of doing the weather was a hit and cross-town rival WTVJ offered him the lead meteorology job in 1989. "They wanted to do something different and dramatic," he said. Translation: the station was willing to shell out big bucks for a fancy weather center, decked out with all the latest gadgets and gizmos.

Beyond the glitz of a new weather center, WTVJ also made an investment that turned out to be a life-saver three years later—they installed several back-up systems that would enable Norcross to broadcast in an emergency weather situation. In 1992, such a situation happened and it had a name: Andrew.

When that Category Four hurricane tore through South Florida, nearly every other TV station in town was knocked off the air. Even worse, the Miami radar was blown out to the Gulf of Mexico. Yet there was Bryan Norcross on WTVJ, continuously broadcasting for 23 hours from a reinforced bunker at the station. His reports on the storm (aided by a back-up connection to the radar in nearby West Palm Beach) were the only source of solace for a terrified populace.

After the storm blew by and the extent of the devastation was beginning to be fully realized, news helicopters showed roofs with the scrawled message "Norcross for President." The adulation culminated in the mayor of Miami declaring September 10, 1992, as "Bryan Norcross Day," and the governor appointed him to a several-state commission on disaster preparedness.

"I was busy nearly every minute of the day," Bryan said, reflecting on his fifteen minutes of fame after Andrew. "It was quite a ride."

12 Ed Phillips
Phoenix, Arizona
KNXV-TV

"Sure it's hot," the old saw goes about Phoenix's weather. "But it's a dry heat."

If you've spent any time in the Phoenix area during the summer, you know this saying is a cruel hoax. According to meteorologist Ed Phillips, the summer in Phoenix can be downright muggy, thanks to a strange quirk in the area's climate called the "monsoon."

And Ed should know—as the area's most popular TV weathercaster, he's seen quite a few monsoon seasons in the nearly 20 years he's spent forecasting the weather in Arizona. Ed explains that the monsoon sets up in July and can run through September. What causes a "monsoon" in a region generally associated with deserts, cactus, and scorching heat? Actually, the heat is one of the factors.

"June is our hottest month," Ed says, noting that temperatures can soar to 110°F and even 120°F. This extreme heat creates an area of low pressure (sometimes called a thermal low). After baking for weeks, this low pressure gets strong enough to act like a vacuum cleaner—it sucks up moisture from the Gulfs of California and Mexico. Sometimes the remnants of a tropical storm that's slammed into Mexico's western coast adds to the moisture flow. The result? A bizarre rainy season from July through September with wave after wave of thunderstorms—and high humidity that can send dew points into the 60s and 70s . . . about the same as sultry areas of the South.

Trying to forecast the monsoon can be tricky; Phillips explains that every day of August will see a chance of showers. Unfortunately, it will only rain one out of ten of those days. In a strange way, this quirky weather fits the funky personality of the state. Interestingly, Phillips' varied career is almost as surprising as the Arizona weather.

Ed Phillips decided to move to Arizona in 1976 after visiting the area on vacation. He had a degree in meteorology from Parks College of St. Louis University but no job offer. After failing to find work forecasting the weather, he took a job at a retail tobacco shop.

Ed's big break into the media came in a rather strange way: one day, a customer with a deep, booming voice walked into the shop. Ever the observant scientist, Phillips asked him if he worked in radio. Yes, the local disc jockey replied, and after talking for a while, he agreed to introduce Phillips to the news director at local radio station KOY-AM.

The rest is Phoenix weathercasting history. Phillips got a job at KOY-AM and worked there forecasting the weather for five years. In 1979, a news director at KPNX-TV called out of the blue and asked Phillips if he wanted to do the weekend weather. "Sure," he said, describing his first television experiences as "terrifying. I was scared to death for the first two years."

Ed worked for two Phoenix-area TV stations (KPNX and KSAZ) during the 1980s before leaving to join upstart KNXV in 1994. "This station didn't even have a newscast before I arrived," Ed said, adding that building a weather center from the ground up "has been quite a ride." Weather buffs praise Phillips' "to the point" presentation with "the best graphics" and "the least-useless information."

Part of Phillips' popularity may be thanks to his weather almanac, which features answers to all of the questions people have asked him about Arizona weather over the years.

First published in 1980, the booklet is the longest continually published local weather almanac in the U.S.; over 100,000 copies are distributed free each year.

"West of the Rockies, you have to throw out all the rules," Phillips says. Thunderstorms drift to the west or southwest—the opposite of the rest of the country. Another anomaly: cold fronts tend to go through dry, with not a shower in sight. And just a two-hour drive from the desert is snow country, where towns like Flagstaff can see significant dumps of the white stuff in the winter.

Speaking of flakes, no biography of Ed Phillips would be complete without mention of his *other* career—in 1990, Phillips was elected to the Arizona State Senate. "I guess folks figured I would make a good politician," he said, pausing for effect, "since I'm used to being half right."

13 Steve Pool
Seattle, Washington
KOMO-TV

"I like to give the forecast in English," Steve Pool once told a newspaper reporter. Pool thinks the problem with most TV weathercasts is that the other guys "tend to get lost in gobbledy-gook. When they're finished, you still don't know if you need an umbrella."

In a town where umbrellas are standard body equipment, Steve Pool has established himself as the preeminent weathercaster—a rather amazing feat considering he fell into TV weather almost by accident.

"I was more heading more towards law in college," said Pool, who attended the University of Seattle after growing up "all over the world," thanks to a dad in the service. Pool's law career was derailed one night when he visited a friend who was doing a show on the college radio station. "I would play various characters on the show, and I had a great time," he said. "I definitely got the bug for broadcasting.

After graduating in 1976 with a degree in broadcast journalism, Pool landed an internship at local TV station KOMO—and has been there ever since. "I started as a production assistant and wrote the scripts for the anchors. Every time they tripped over a word, I'd wince, thinking it was my fault."

Pool's career has metamorphosed over the years at KOMO, with the broadcaster doing turns as a reporter, sports anchor, and talk show host before he happened upon the weather in 1983. KOMO's long-time weathercaster Ray

Ramsey went on vacation that summer, and the station asked Pool to fill in.

"I had a blast," Pool said, adding that the unscripted and ad-libbed weather segment reminded him of his free-wheeling days on college radio. In order to know what he was talking about, the station sent him back to the University of Washington to study meteorology for a year. While he didn't earn a second degree, he took enough courses in atmospheric sciences to prepare him to forecast the Pacific Northwest's wild weather.

"In the space of 250 miles around Seattle, you have an ocean, an inland sea, two mountain ranges, a desert, and a rain forest," Pool said. The result is a weather pattern that doesn't fit neatly into the weather service's computer models.

An example: the "Puget Sound Convergence Zone." When a storm slams into the area, the Olympic Mountains to the west of Seattle can split the storm literally in half. Then, the storm reforms, "converging" in the Puget Sound area with howling winds and rain. The result: 20 miles to the west of Seattle, it can be sunny and warm, while the city itself is drowning in rain.

"And here's the tricky part," Pool explains further. "The zone itself can move, throwing everything out of whack." The area's quirky microclimates require Pool to draw up maps with precision. He also relies on an extensive network of weather watchers to feed the station with daily reports.

"Sure, Seattle has a reputation as a rainy city, but, hey, we only get 36 inches of rain per year," Pool said. That's less than cities like New York and Washington, D.C. The rub is the rain doesn't fall as rapidly as it does on the East Coast; Seattle can be bathed in drizzle for days.

"This city has a relatively regular weather pattern. It rains from February to July 6, then summer begins," Pool said. This lasts until October, when the clouds roll in. Kiss the sun goodbye until next July.

14 Tom Skilling
Chicago, Illinois
WGN-TV

Tom Skilling is a TV weather god. There probably isn't another weathercaster in this country (or the planet, for that matter) who combines a fanatical enthusiasm for the weather with a potent, rapid-fire graphics presentation than Skilling.

And that isn't half bad for a guy who once was forced to do a weather report with a puppet named "Albert the Alley Cat."

Before we get to the cat story, I should note that Tom grew up as a serious weather enthusiast whose broadcast career started when he was just 14. In a recent phone interview, he traced his phenomenal weather career.

"I had paper routes to raise the money to buy weather instruments," said Skilling, who was born in Pittsburgh but grew up in the Chicago suburb of Aurora. "My dream was to have a weather radar set-up in my bedroom. I actually went as far as writing to the company that makes the radars for brochures."

At the age of 14, he had the audacity (his words) to approach local radio station WKKD about doing the weather. Tom thought he could do a better job forecasting for Aurora than the reports the station was getting from Chicago. To his surprise, they hired him. "I suppose they thought it was quite a novelty to have a local fourteen-year-old boy doing the weather."

So, Tom started his broadcast career, phoning in his forecast to the station before he headed off to high school. After doing forecasts for WKKD for three years, Skilling was hired to do the weather for a new local television station (WLXT-TV) at the tender age of 18. By the time he headed off to the University of Wisconsin at Madison to study the weather, he already had four years of broadcasting under his belt.

Tom's college years were filled with more broadcast jobs, including doing the weekend weather at a local station (WKOW), forecasts for WTSO-Radio, and even private forecasting work for a brokerage firm. You could say he majored in work and went to school in his spare time. As a result, he never got a degree and instead took a weather job with an NBC affiliate in Jacksonville, Florida, in 1974.

Within a year, Tom was offered a job with ABC affiliate WITI in Milwaukee, and anxious to return to the Midwest, he took it—and that's where the cat comes in. You see, prior to Tom's arrival, the weather at WITI was done by a puppet, Albert the Alley Cat. Even though Albert was incredibly popular, the station promised Tom that they would be phasing out the cat as their chief weathercaster.

Three years later, it was Tom who was sent packing and the cat who remained.

The whole ruckus started when the station tried to phase out the aforementioned puppet. Call it the Revenge of Albert—the station started receiving sacks of hate mail demanding that the puppet be restored to his rightful role as the weathercaster on WITI.

"So, here I was, Mr. Scientific with all my jet stream maps and isobars, paired with a puppet who did the current conditions," Skilling recalls. It would have been more funny if the American Meteorological Society (AMS) hadn't called one day.

"They told the station that if they didn't get rid of the cat, they would strip me of my AMS seal of approval." The ensuing controversy got ugly when word of this hit the local papers. At the same time, the station's consultants were criticizing Tom for being "too scientific." When the station began editing his maps, telling him to drop the jet stream stuff, it was time to get out.

Tom wasn't unemployed for long. WGN in Chicago called him in for an on-air audition and soon offered him a "dream weather job"—doing the weather in prime-time for one of the biggest markets in the country. Yet despite all his experience, Skilling had self-doubts and wondered whether, at the age of 25, he was ready for such a big market.

"My friend, John Coleman (of "Good Morning America" fame and the founder of The Weather Channel) told me I was crazy. An opening like this at a major television station might only happen once every ten years—he thought I would be nuts to pass it up."

The rest is history, as the saying goes. Tom Skilling began the process of updating WGN's weather office into one of the most technologically-advanced centers in the country. "When I started here, we had no computers," he recalls. "Weather maps were drawn on Plexiglas boards."

By contrast, Tom's weathercast today is a high-tech marvel—it takes him twelve to thirteen hours each day to put together all the animated jet stream maps and forecast graphics. All this work is rewarded by WGN's huge audience via cable TV; the station is seen by 40% of all households in the U.S. with cable—that's 37 million homes.

"I get letters from Brazil, Hawaii, and Alaska," Tom said. Even though his forecast and maps center on the Chicago area, Skilling thinks his "big-picture" approach to presenting the weather explains why he has so many fans outside of the Windy City. "It amazes me how many write in to say how

WGN is much better than their local weathercast. It makes me wonder what's going on out there."

15 Mike Thompson
Kansas City, Missouri
WDAF-TV

Mike Thompson's first television weathercast was to a very small group of people who listened intently to every word he said. "I got my start as the meteorologist on the aircraft carrier USS *Lexington*," Mike said, who went through the naval meteorology college in Lakehurst, New Jersey.

Through closed-circuit broadcasts, he briefed the captain and pilots on weather conditions in the Gulf of Mexico, where the carrier often sailed on training missions.

After leaving the Navy in 1979, Mike drove back to his hometown in Burlington, Kansas, and applied for any open meteorology position, hoping to land a government job. "TV never crossed my mind," said Mike, who credits his upbringing in tornado alley for his interest in the weather.

Mike eventually landed a job in Oklahoma City, working for a private forecasting company. One day, his boss asked him if he'd thought about doing television, noting that a new UHF station in town (KOKH) was looking for a weathercaster. And that's how he launched his weather career, spending six months there before jumping to KWTV, the CBS affiliate in town. After a brief stint in Charlotte, North Carolina, he moved to Kansas City in 1982 to work for the CBS affiliate KCMO, now KCTV. He's been in Kansas City ever since, moving to the Fox affiliate WDAF in 1992.

Mike's passion is severe weather, and Kansas City sure sees its fair share firing up near the metro area. "We seem to be at a turning point for air masses," he commented, noting that gulf moisture flows into the area and makes a right turn at the city. In the winter, Canadian air masses slip just to the west. Kansas City is like the nation's weather cop, in the middle of weather "traffic" but always just one step from the action. "Just when you expect rain or thunderstorms to form, they won't for that reason."

One particularly vexing forecasting dilemma: the summer thunderstorm that forms in Kansas and moves straight for Kansas City—only to turn south at the last minute and miss the city. "The old school of thought on why this happened was the mid-level winds (at 10,000 feet) were steering the storms away from us," he said. "Today, we think the storms move that way because weak upper-air winds let the storms follow their low-level feed of moisture"—this southerly wind flow literally sucks the storms away from Kansas City. The technical term for these storms are "mesoscale convective complexes," huge clusters of thunderstorms that seem to create their own weather.

Of course, it isn't as though Kansas City *never* sees a thunderstorm. The area has experienced its fair share of heavy weather, including that particularly eerie phenomenon that accompanies severe storms on the Great Plains: the "green sky." Mike explained what makes a normally black storm cloud take on a bizarre green hue.

"It usually happens late in the day," he said. "The green color comes from the sun shining through hail—the stones of ice act like a prism and scatter more of the green light than any other color. The angle of the sun is also critical—if the sun is low in the sky, the light is already distorted. The hail stones further distort the sunlight into an eerie green glow."

3

Weather You Can Always Turn To: Backstage at The Weather Channel

I T'S 7:20 PM ON MAY 2, 1982. In just 40 minutes, the first 24-hour weather cable channel will sign on the air . . . and Alan Galumbeck is in a panic.

Galumbeck, the network's designated computer whiz, has just discovered a bug in the computer system that broadcasts the local forecast, time, and temperature to affiliates. As a result, The Weather Channel is about to debut with the incorrect time splashed across television screens in the central, mountain and Pacific time zones.

Big deal, you might say. Since just 3.2 million households will be able to see the network's debut, there must be only, say, 57 people in the Pacific time zone who would even see the goof. Unfortunately for The Weather Channel, those 57 people just happen to be the heaviest of heavy hitters in the fledgling cable industry.

"We had a big launch ceremony planned for a cable industry convention in Las Vegas that night," Galumbeck explains twelve years later. Dozens of cable industry executives and affiliates were huddled around television screens, anxious to see the first all-weather channel premiere its maps and forecasts. And the correct time would be a nice bonus.

For months, The Weather Channel had been touting the superior computer technology that would enable it to broadcast each of its affiliates' local forecast. The same computer would also scroll the current time and weather conditions at the bottom of the screen during the channel's regular weather segments. Local cable companies insisted The Weather Channel regularly broadcast local weather conditions and forecasts—or else, the channel would be dropped faster than you can say "cumulonimbus."

Galumbeck had been testing the computer system (called the Weather Star) for weeks. Unfortunately, the Weather Star had been showing only the correct eastern time, and since the network's offices were in Atlanta, the error went undetected.

Working furiously to reprogram the Weather Star, Galumbeck managed to get the computer to display the correct time in the Pacific time zone at 7:55 PM—just five minutes before the signal was beamed to the cable convention in Las Vegas.

Disaster was averted. The affiliates cheered. The meteorologists breathed a sigh of relief. Yet The Weather Channel's journey from obscure cable network to the most trusted source for weather information was just beginning.

Weather and Your Pork:
The Early Years of Bac-Os and Losses

John Coleman is the weatherman credited for coming up with the idea for a 24-hour weather channel. After founding The Weather Channel, Coleman went on to "Good Morning America" and is still weathercasting today at a station in Palm Springs, California.

While the enthusiasm of meteorologists like Coleman certainly helped the network, it was the deep pockets of Landmark Communications (a media company that owns several newspapers, television stations, and other businesses) that saw the network through those dark early days.

How dark? The Weather Channel reportedly lost $800,000 a month during 1982 and 1983. Critics dismissed the channel as a joke and took bets on how long it would be before the plug was pulled. Convincing advertisers that anyone was watching became one of The Weather Channel's biggest obstacles to profitability. Because its subscriber base was so small in 1982, ratings companies such as A.C. Nielsen wouldn't even count the network's audience. Without any official ratings, The Weather Channel had a devil of a time attracting advertisers.

The network's desperation for revenue apparently drove it to promise advertisers additional "infomercial" segments that shamelessly plugged their products. One such feature was called "Weather and Your Health," sponsored by Bac-Os, that artificial bacon condiment made by the authority on healthy living, Betty Crocker. These segments featured one of the on-camera meteorologists extolling the benefits of a healthy diet. "One good-for-you alternative is a salad of fresh greens," said the weathercaster, who looked like she'd prefer getting a root canal to doing this commercial plug. And what would be the perfect topper for that healthy salad? Bac-Os, of course. The segment featured a shot of the product on the kitchen counter and the weathercaster sprinkling the infamous imitation-bacon product on the aforementioned salad. Right after "Weather and Your Health" was a commercial for, surprise, Bac-Os. Perhaps this segment should have been renamed "Weather and Your Pork."

But it wasn't Bac-Os, Salad Shooters, or Time-Life's "Hits of the 1920s" that saved the Weather Channel. No, it was Bob. Or more precisely, a hurricane named Bob.

When Hurricane Bob threatened North Carolina, a funny

thing happened: The Weather Channel's viewership soared—and big-time advertisers soon followed. And the same phenomenon still occurs today when a major storm blows toward the coast. The fact is, while The Weather Channel is available today in over 55 million households, only 100,000 or so homes are watching at any one time. Yet when a storm like the Blizzard of March 1993 occurs, The Weather Channel's ratings jump tenfold. 1.2 million homes were tuned in the Saturday evening during that famous blizzard—that's more than CNN's audience at that time.

Backstage at the Weather War Room

The Weather Channel has come a long way since the days of Bac-Os and Hurricane Bob. The polished and informative presentation of the on-camera meteorologists has created legions of Weather Channel addicts—myself among them.

Despite this success, however, the Weather Channel doesn't offer public tours of its studios . . . and it's probably a good thing. You might expect the network to have offices in a downtown skyscraper in Atlanta, perhaps right across the street from CNN. Yet instead of a penthouse suite with sweeping views of clouds and the city, the meteorologists are stuck in suburban Atlanta with a garden-level office in a building that looks like a bomb shelter. The single, solitary window in the "Forecast Center" (the computer area you see behind the on-camera meteorologists) looks out on a parking lot. And even that window was only recently added when a visiting reporter noted it was ironic that the country's all-weather network didn't have a way for its meteorologists to look outside.

Still, The Weather Channel's suburban location is perhaps fitting for a network whose on-camera weathercasters

41

are nearly all white men. Oh sure, there is one African-American meteorologist, and several of the women weather-casters have quite a large following—but if you turn on The Weather Channel at any random time, odds are the weather is being delivered by an Anglo male.

The suburban digs also seem appropriate given the mete-orologists' earnestness. This isn't hip, urban "Weather TV" where quick cuts of flashy graphics are interspersed with jokes and light banter. The Weather Channel takes its weath-er seriously, thank you. While this presentation of the weather is certainly more watchable than the Ronald McDonald School of Weather Jokesters, you can't help but wonder if the on-camera meteorologists' delivery of winter storm forecasts and tornado warnings would be more pas-sionate if they were actually closer to what was going on out in the real world.

Para-social Scores: Para-normal or Para-strange?

With all the high-tech Doppler radars and satellite pic-tures that permeate the halls of The Weather Channel, you might forget that this is actually television, not science. In case you need a dose of reality, drop by Craig Brewick's office.

Brewick supervises the on-camera meteorologists and is, as you might guess, not a meteorologist but a television exec-utive who's served as the news director at a couple of big ABC affiliates. Surrounded by an office full of thick research binders, Brewick is steeped in the latest TV jargon, such as how the weathercasters rate in "para-social scores." As one media critic once wrote, para-social scores are an "anti-septic term which TV executives use when they talk about making their news teams part of your family."

Perhaps the only thing more absurd than ranking meteo-rologists on their "para-social ability" is the fact that network executives actually believe this stuff. Brewick earnestly explained to me how important this is, as if people not only want their weekend forecast from weatherman Jim Cantore but would also like to have him over for lunch.

And that's just the beginning—another criterion Brewick uses to evaluate the meteorologists is how often they men-tion the "big cities." The more the weathercasters mention big markets like New York, Chicago, or Los Angeles, the more brownie points they earn. New York City may be home to ten million people, but in the big picture, that's less than 5% of the country's total population. While the entire popu-lation of Montana is dwarfed by the Bronx, the discussion of weather shouldn't be focused on areas with the biggest TV market rankings. I always wondered why sparsely populated areas of the country got short shrift on The Weather Channel, and now I know why.

Next, Brewick looks at "map blockage," the ability of the

meteorologists to give the weather without their fanny blocking out all of North Carolina. When the publicist who was accompanying me on the interview told Brewick that The Weather Channel had received no complaints about "map blockage" that week, Brewick broke into a big smile. I suppose to him this is a big deal, since a wayward facial feature such as a big nose might knock dinky states like Rhode Island or Delaware into weather oblivion.

One of the more interesting evaluation criteria is the meteorologist's "story telling" ability. The weather is like a novel, explains Brewick, with a plot and characters that all need explaining. While anyone can read a weather map, weaving those cold fronts and squall lines into a compelling "story" is apparently what separates the weather*men* from the weather*boys*.

When our conversation turned to money and how much The Weather Channel pays its meteorologists, Brewick gave a "no comment," except to admit that the salary packages are less than what's being offered to weathercasters by big-market TV stations. "We offer more quality of life," he said, adding that if the weathercaster is really into the weather, there isn't a place that has more toys than The Weather Channel. Besides, the network offers the chance to report on severe weather every day of the year, in contrast to boring weather cities that might see two months of action followed by weeks of static weather.

It's clear that the Weather Channel's mostly white staff is partially a result of the salaries (or lack thereof). Minority weathercasters are hot properties since TV stations seek to give their Anglo-heavy news teams more "ethnic diversity." The Weather Channel's reluctance to pay the going rate to bring an Asian or Hispanic on board is out of step with the times. While Brewick told me he's trying to recruit more minorities to be on-camera meteorologists, the jury is still out on whether the Weather Channel will look like America in the 1990s anytime soon.

The 11:30 Briefing

After the meeting with Brewick, it was comforting to step back into the real world, back to the weather. I sat in on the 11:30 "weather briefing," one of three meetings held each day to update the staff on changing conditions. These businesslike meetings don't include much idle chatter—the staff meteorologists who toil behind the scenes give the weathercasters who are about to go on the air a quick run-down of what's happening with the weather. The meteorologists sit in front a big TV and review many of the maps and satellite images that will be used in upcoming broadcasts.

This July morning, an unusually strong cold front has brought record lows to the South; otherwise, it's a rather hohum weather day. After John Hope, The Weather Channel's

"Tropical Coordinator," gives the weathercasters an update on the Tropics, the meeting is over. Total time to review the current conditions and forecasts for the United States and Europe: 10 minutes.

Waving at the Air

The on-air studio is located right next to the "Forecast Center." Although on TV the two areas appear to be connected, they're actually separated by a sheet of Plexiglas. The studio has a desk at which the meteorologists sit during the introduction of each weather segment. On either side of the desk is a "chroma key" area, a blue screen in front of which the weathercasters stand when giving the weather—the actual weather maps are superimposed on the blue screen. In order to see where they are pointing, the weathercasters must look at a monitor just off camera. No matter how many times I have seen it before, I still find it eerie to watch a weathercaster gesture at a weather map that really isn't there. The illusion is even more remarkable in person—even when they look at the monitor, you still wonder how they know they're pointing at Vermont instead of Maine.

Even more amazing is how the weathercasters correctly pronounce the names of all those small towns in Vermont, or in any state for that matter. Keith Westerlage, an on-camera meteorologist who's been at The Weather Channel since 1987, revealed to me the secret: they cheat. By using a specialized geography dictionary (for the record, *Webster's New Geographical Directory*), they know that Louisville, Colorado, isn't pronounced the same way as, say, Louisville, Kentucky. And when severe weather threatens a place like Austin, Texas, you'll hear the meteorologist flawlessly spit out town names like Manor (pronounced May-nor) and Pflugerville (the "p" is silent). Keith told me that some of the meteorologists can just look at the dictionary once and then regurgitate the towns and pronunciations by memory—others (like him) cheat by drawing a small map of the towns and posting it just out of camera-range.

Brutal Schedules

One aspect of The Weather Channel that surprised me was the brutal schedules of the on-camera meteorologists. Marny Stanier's day is typical. A popular seven-year veteran of The Weather Channel, Stanier starts her day at 3:30 AM. By 4:30 she's at the office, and she'll do the weather (on alternating half-hours) until 1 PM. On the upside, she only works four days a week and gets Friday, Saturday, and Sunday off.

Other weathercasters aren't so lucky—they may put in eight or nine hours during the graveyard shift, working weekends and holidays. During severe weather outbreaks, the entire staff may pull an all-nighter. A hurricane turns the place upside down—updates are done "live" from the Forecast Center, and crews are sent out to do remote broadcasts from likely landfall spots.

If I Programmed The Weather Channel . . .

Let me state right here and now that I think The Weather Channel is the best invention since sliced bread. As an avid viewer, however, I believe it's my prerogative to pontificate on how to make The Weather Channel better—like the armchair football fan who thinks he can call the plays better than the head coach.

So, here are my five suggestions for improving The Weather Channel:

1. BEEF UP THOSE WEATHER MAPS. The Weather Channel has been somewhat slow in adopting the latest technology in three-dimensional maps and animated weather graphics. Its flat two-dimensional maps look dowdy compared to the snazzy 3-D effects on local weathercasts. While you can argue whether improved graphics (3-D, animation, etc.) increase understanding of the weather or are merely a gimmick, it's rather obvious that The Weather Channel is more of a follower than a leader in technology. It should be the other way around.

Another missing map element: topography. Most of The Weather Channel maps have no topographical information. Pinpointing foothills, mountains, and other topographical features makes a world of difference in understanding the weather. For example, in the western U.S., counties can go from 5000 feet in elevation to 13,000 in a matter of a *dozen miles*. Highways would also make a big difference—most viewers know their location relative to the nearest highway, not a dot on a map that says Kansas City.

The Weather Channel told me they're making improvements in this area—a new mapping software program is being tested that will pinpoint suburbs, landmarks, and other features on radar maps.

2. STOP THE "PING-PONG FORECASTS." When you watch The Weather Channel, you'll notice your local forecast pop up on the screen several times an hour. In an amazing technical feat, the network delivers this forecast to each of its cable affiliates, who in turn actually broadcast it to your TV. Unfortunately, while The Weather Channel does an admirable job of passing along severe weather information, its forecasts can drive me nuts.

Now, I should note that I've got no problem with the 36-hour forecast (the script part of the forecast that reads "today. . . partly cloudy with a high in the 70s"), which comes directly from the National Weather Service. No, what's got me in a tizzy is the long-range or five-day forecast, which is actually determined by the Weather Channel, not the National Weather Service. A computer looks at the long-range maps prepared by the station's meteorologists and predicts the weather and temperatures for a particular city. The problem? I call it the ping-pong forecast syndrome—one hour they'll forecast certain temperatures, only to change them dramatically in the next hour.

This problem is most pronounced in the winter, when a large outbreak of cold air from Canada threatens to send the country into a deep freeze. Certainly, any long-range forecast is tricky—a slight movement in the direction of the cold front could mean the difference between a sunny, warm day and a blizzard. Yet, it drives me to distraction to see the temperatures ping pong around from one hour to the next. For example, a major cold front threatened the Front Range of the Rockies during one week last winter. When I turned on my TV Monday morning (about three days out from frontal passage), The Weather Channel was forecasting highs in the 30s for Thursday. By noon, Thursday's high temperature forecast had dropped into the single digits. Then another revision came that afternoon, when the temperature forecast popped back into the 20s. Ping. Pong. Ping. More care and thought needs to go into these forecasts to keep from confusing viewers.

3. LONGER BLOCKS OF WEATHERCASTS. The constant shuffling of weather maps on The Weather Channel certainly enlivens their broadcasts, but sometimes I wonder if the breakneck pace (with frequent commercial breaks) is doing a disservice to the viewer. It'd be nice if The Weather Channel would give me some time to really study a satellite picture or radar image.

I should note that The Weather Channel is currently expanding to longer blocks of weather information at the top of the hour during prime time (7 to 10 PM eastern time). These longer broadcasts go into more detail than the usual quick round-up, and I hope they become a fixture of the network in the future.

4. MORE RADAR. The latest high-tech radar (NEXRAD) can provide an incredible amount of information to weathercasters, including wind direction profiles (important in predicting possible tornadoes) and even rainfall rates (interesting when a tropical storm makes landfall and threatens to soak a state like, say, Georgia).

Despite the enormous potential of this weather tool, The Weather Channel uses little of the expanded NEXRAD capabilities. If anything, the network tends to use NEXRAD as "after the fact graphics," for example, showing how the radar captured a picture of a tornado *after* it struck. I'd like to see more use of this important technology to illustrate what's going on now, not yesterday.

5. LIVE SHOTS FROM DIFFERENT CITIES. The Weather Channel should set up live cameras in major metro cities to report what's happening right now. Local stations have installed automatic "live action cams" on poles to create time lapses of the day's weather. The results are spectacular: in a montage that lasts just a few seconds, you can see clouds boil up, storms breeze by, and more.

Sure, it's one thing to look at a storm on the radar and ponder it on the satellite. But to actually see the storm cloud towering over the horizon or watch "live" a vicious squall line adds a unique perspective. The Weather Channel should set up its own network of live cameras or at the very least, tap into the live pictures from existing local station's cameras.

The Local Forecast Soundtrack Album

What's the number one question The Weather Channel gets from its viewers? If you guessed something to do with meteorology, guess again.

The thing viewers want to know most is: what is that darn music that's played in the background during the local forecast?

To find the answer, I visited Chris Hoitsma, "viewer services coordinator" for The Weather Channel. In his cubicle downstairs from the studio, he fields dozens of calls and letters from viewers. There have been so many inquiries about the music that Chris has created an official list of all the songs played during the local forecast. In case you're curious, here's a rundown of the artists that make the local forecast segment so soothing:

> • *3rd Force*. 3rd Force is probably the artist played the most on The Weather Channel. The network uses snipets from seven new age songs from their self-titled debut album (the publisher is Aura Communication), including "We Should be Together," "You Know My Heart," "Ready for A Cold Front." (Just kidding on that last one.)
> • *Shadowfax.* Another popular new age artist, Shadowfax places several of its songs on The Weather Channel's hit parade. Tune into the local forecast and you'll hear parts of "The Return of the Nairobi Trio," "Found Wind," "Tropico Blue," and "Blue in the Face" (from the *Esperanto* album), as well as "Baker's Dozen" (from the *Magic Theater* album).
> • *Gary Brunotte*. The Weather Channel uses four cuts from Gary's *Yesterday's Dream*, including "Slightly Blued" and "Caterpillar Crossing." The publisher of these songs is Summit Music.
> • *And more.* Several artists are used less sparingly on The Weather Channel. The roster includes After Five Jazz ("My Latin Lady" from the *Expressions* album), Steve Smith ("As the Crow Flies" from the *Distant Lands*

album), EKO ("Midnight Panic" from the *Alter* album), and Shahin & Sepehr ("Yasmine" from the *1001 Nights* album).

Merchandising the Weather

While the Weather Channel hasn't announced plans for a CD soundtrack of the music played during their local forecasts, the network does sell a variety of weather instruments, clothing, and sports items—all emblazoned with their familiar blue logo.

You can get a "cotton deluxe turtleneck" with the Weather Channel logo for $12.83. A baseball cap comes in adult, toddler, and infant sizes ($5.78 to $6.60).

Linksters can buy The Weather Channel golf balls ($18.95), a golf umbrella ($17.99), or even a clear water bottle ($1.70). Relax after the game with a Weather Channel beverage koozie (95¢) and add up your score with a Weather Channel "glitter brite pencil" (27¢) or Bic pen (65¢).

Of course, The Weather Channel also sells various items for weather buffs. The network's official rain gauge is $21.38, and a patio thermometer is $8.25. There is even a Weather Channel ice scraper (95¢) and umbrella ($9.65).

All of these products can be ordered by calling (800) 537-6765 or (404) 552-9127. A catalog is available.

In addition to exciting merchandise, The Weather Channel also offers videos of the special programs and documentaries that have recently aired on the network. The price per video is $19.95 plus $3.95 shipping. Unfortunately, each video has a different 800 order number. Here's a rundown of the videos available, including their order numbers:

"The Enemy Wind" (800) 525-6600
"The Burning Season" (800) 538-0505
"The Year the Sky Fell" (800) 841-4545
"Out of the Blue" (800) 385-3344

Forecasting the Future of The Weather Channel

With all the talk of the coming "information superhighway," what lies ahead for The Weather Channel in the 500-channel, interactive video future of cable?

To find the answer, I spoke with Alan Galumbeck, The Weather Channel's point man for future projects. Yes, that's the same Alan Galumbeck who was furiously working to reprogram the network's computers before its launch 12 years ago. Today, Alan's official title is "Senior Vice President, New Technologies" for Landmark Video Networks and Enterprises (the parent company of The Weather Channel).

Galumbeck told me The Weather Channel of the future may rely heavily on the emerging "video on demand" technology. Basically, you will be able to pick which segments you want and then watch them when you're ready. Assume you're going skiing—you will be able to request such segments as the eastern U.S. ski report, your local weather, and the five-day business planner. With a touch of a button, the most recent of these segments will be zapped to your TV, displaying graphical information plus "full-motion" video of a meteorologist doing a report on the ski conditions. Each viewer would become his or her own program director for The Weather Channel. What if you want to be a couch potato? Instead of picking what segments you want to see, you could access a "standard profile" of various weather reports, much like The Weather Channel is programmed today.

One interesting side effect of this "interactive video on demand" future may be longer and more in-depth weather reports. Right now, The Weather Channel has to break frequently to broadcast "local forecasts," one of the most viewed features. In the future, viewers who just want this information can bypass the more lengthy weather reports and simply pop up the local forecast on their TV. According to Galumbeck, The Weather Channel may be able to do longer reports on the "whys of the weather," and these segments would contain much more in-depth information on how the weather works.

While video on demand is probably ten or more years in the future, The Weather Channel is proceeding with a new project that may debut as soon as next year: on-line weather. The network is negotiating with several of the major on-line services to provide a "Weather Channel Forum." You'd be able to view specialized maps on your computer, as well as converse with the network's meteorologists and even download pictures of your favorite weathercasters. Another posting in The Weather Channel Forum may answer that burning question in many viewers' minds—what are the songs being played during the local forecast segment? See page 47 earlier in this chapter for the answer.

Screen Phones to the Rescue

What if you don't have a personal computer? Galumbeck is also hard at work on another future project involving "screen phones." Now, these are not video phones—a screen phone doesn't show any pictures but instead displays four to six lines of text. This technology isn't some pie in the sky—screen phones will be on the market in 1995.

How will The Weather Channel use screen phones? The network sees this as a means to augment its 1-900-WEATHER service—screen phones would let you navigate the menus faster, among other benefits. In addition, the hearing-impaired would also be able to get weather information by screen phone.

The Talking Local Forecast

The Weather Channel is also looking to beef up its local forecast segment. Today, you have to read the forecast text off the screen—in the not so distant future, you may have the option of having the TV read it for you. "Voice synthesis" is a possible new feature for the Weather Star, the computer that helps local cable affiliates display the forecast information on your TV.

Besides voice forecasts, look for better graphics during the local forecast segment on The Weather Channel. Galumbeck said the next generation Weather Star will be capable of showing topographical maps, zoom in on local radar loops, and more.

And there's a decent chance the correct Pacific time will be displayed, Galumbeck added.

Show & Tell: The Secret Lives of the Weather Channel Meteorologists

W HICH WEATHER CHANNEL METEOROLOGIST was a world-class track athlete in college? Who was known as "Weather Will" when he worked in Green Bay, Wisconsin? Which weathercaster teamed with his wife on a local TV station to become the first husband and wife weather couple in the U.S.? Who lists "aerobics" as a hobby? Which meteorologist spends his spare time creating greeting cards?

Here are the answers, along with biographies on the 30 meteorologists who make up the on-air staff at The Weather Channel.

★ Will Annen

Current conditions: Married with two daughters. Lives in Marietta, Georgia.
Hometown: Madison, Wisconsin.
College: University of Wisconsin, bachelor's degree in meteorology, 1977.
Hired by The Weather Channel: 1982. He was there at the beginning and has stayed at the network ever since.
Quirky nickname: Weather Will, bestowed on him by the Great Lakes Weather Service, where he worked in the late 1970s. When he moved to TV at WLUK in Green Bay, the name stuck.

★ Mike Bono

Current conditions: Married with one son.
Hometown: Bayside, New York.
College: New York University, bachelor's and master's degrees in meteorology, 1971 and 1975, respectively.
Career Track: Started at WLBZ in Bangor, Maine, before landing a spot on a cable TV network in Long Island. Landed at The Weather Channel in 1987.

★ Jill Brown

Hometown: Wooster, Ohio.
College: Ohio State, B.S. in Aviation Engineering, 1986.
Career Track: Started her career at The Weather Channel doing graphics and radio broadcasts. Left to be the morning weather anchor and environmental reporter in Hartford, Connecticut, at WFSB, before returning to the weather network in 1992.
Trivial fact: She's one of the few meteorologists who lists golf as a hobby.

★ Vivian Brown

College: Jackson State University, B.S. in meteorology, 1986. Brown received a full-athletic scholarship, ranked fifth in the world in 50-meter track competition in 1985, and was voted the Most Valuable Player in field events in the Southwestern Athletic Conference.
Career Track: Joined The Weather Channel in 1986, preparing maps and doing behind-the-scenes forecasting work. Brown enrolled in 1988 in the network's apprentice program and became an on-camera meteorologist in 1989.

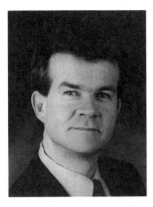

★ Declan Cannon

Hometown: Annapolis, Maryland
College: Lyndon State College (Lyndonville, Vermont), B.S. in meteorology, 1981.
Career Track: Worked for a private forecasting firm for a year before joining The Weather Channel in 1982 as a forecaster. His first on-air broadcast was in 1986.
About His Name: It's Irish.
Style: Soft-spoken but effective.

★ Jim Cantore

Hometown: White River Junction, Vermont
College: Lyndon State College (Lyndonville, Vermont), B.S. in meteorology, 1986.
Career Track: Was a DJ at WHDQ-FM in Claremont, New Hampshire, before jumping into TV as a part-time weathercaster for WCAX-TV in South Burlington, Vermont. Joined The Weather Channel in 1986. *Style:* Popular weathercaster who likes to use his fingers to point out individual thunderstorm cells on the radar.

★ Dale Eck

Hometown: Eaton, Pennsylvania.
College: Penn State, B.S. in meteorology, 1981.
Career Track: Worked with Accu-Weather, forecasting ice and snow conditions for trucking companies. Jumped into TV in Macon, Georgia, at WGXA-TV before joining The Weather Channel in 1986.
Fun with Bios: Eck's official biography says he's "sports-minded and regularly jogs."

★ Brad Edwards

College: Iowa State University, B.S. in meteorology.

Career Track: A recent hire at The Weather Channel in 1993, Edwards began his TV career as the morning and noon meteorologist at WTVX in Ft. Pierce, Florida. In 1988, he left Florida for Kansas, joining KTKA-TV as chief meteorologist in Topeka.

Style: Edwards' super-smooth style is a treat to watch.

★ Rick Griffin

College: Lyndon State College, B.S. in meteorology, 1977.

Career Track: Joined his wife, Vicki, to become the first husband and wife weather team at WKIV-TV in 1984. Griffin started his career at The Weather Channel as an on-camera meteorologist at the network's launch in 1982 before leaving two years later. His TV weather resume includes stations in Detroit, Massachusetts, Rhode Island, and California, before returning to The Weather Channel in 1993.

Style: Griffin's extensive experience shows—he's probably one of the better weathercasters on the network.

★ John Hope

College: University of Illinois, Phi Beta Kappa. University of Chicago, Master's of Science in meteorology,

Career Track: Hope has forgotten more about hurricanes and tropical storms than most weathercasters will ever learn. He must have been present at the creation of tropical weather, having spent 40 years in weather research and forecasting. Before joining The Weather Channel in 1982, he was the Senior Hurricane Specialist at the National Hurricane Center in Miami, Florida.

Style: Hope's unpretentious and science-oriented presentation of the "Tropical Update" on The Weather Channel is one of the network's strengths. His style isn't flashy and really hasn't changed much over the years; he's sort of like the Orville Redenbacher of tropical storms.

★ Rich Johnson

College: Florida State University, graduated in 1983.
Career Track: Rich has spent his entire career at the Weather Channel, first starting as a part-time "forecaster assistant." In 1984, he was promoted to janitor. Just kidding—his next job involved creating map graphics, and he did time in the network's apprentice program before making his on-air debut in 1990. *Style:* Imagine Dave Barry doing the weather.

★ Jeanetta Jones

College: University of Georgia, bachelor's degree in broadcast news, 1982.
Career Track: Even though she's one of the few weathercasters at the network who isn't a meteorologist, Jeanetta Jones is one of The Weather Channel's most popular personalities. She started as a reporter and anchor for WMAZ in Macon, Georgia, and later went on to WSPA-TV in Spartanburg, South Carolina. Jeanetta did weekend weather, reporting, and even hosted a talk show at that station. She joined The Weather Channel in July 1986. *Style:* The Kathleen Turner of weather personalities, she's smooth, knowledgeable, and has some of the nicest hair on cable television.

★ Bill Keneely

Hometown: Wayne, New Jersey.
College: Plymouth State College of the University of New Hampshire, B.S. in atmospheric science, 1979.
Career Track: Keneely is another "lifer" at the network, doing on-camera work at The Weather Channel since its launch in 1982. Before that, he worked as a weekend meteorologist for WTVT-TV in Tampa, Florida, and did radio and private client forecasting.
Style: Sure he's a nerd, but he's professional, damn it.

★ Cheryl Lemke

Hometown: Omaha, Nebraska.
College: Iowa State University, B.S. in meteorology, 1982.
Career Track: Lemke got her start as an environmental reporter in Terre Haute, Indiana, at WTHI-TV. She also spent time in Iowa, doing the weather for stations in Ames and Des Moines. She joined The Weather Channel in 1986.
Strange hobby: Believe it not, Lemke lists "aerobics" as one of her hobbies. I'm sorry, but no one on the planet should consider this activity a hobby.

★ Mark Mancuso

Hometown: West Newton, Massachusetts.
College: Penn State, B.S. in meteorology, 1979.
Career Track: He's been with The Weather Channel since its inception; before that, he did weather stints at TV stations in Tennessee and North Carolina.

★ Thomas Moore

Hometown: Rome, New York.
College: University of Maryland, B.S. in meteorology, 1980.
Career Track: One of the original meteorologists, Moore joined The Weather Channel in 1982. Before that, his previous TV experience could be described as "sparse"—he did the weather on a cable station in Oswego, New York.
Turn-Ons: Lake effect snow.
Fun with Bios: Moore lists his hobbies as "outdoor spectator sports and visiting historical sites."
Style: A cross between John Madden and Rip Torn.

★ Jeff Morrow

Hometown: Aliquippa, Pennsylvania.
College: Penn State, B.S. in meteorology, 1980.
Career Track: Before landing at The Weather Channel in 1985, Morrow was a forecaster and "radio personality" for Accu-Weather in State College, Pennsylvania.
Style: Entertaining, yet informative.

★ Jodi Saeland

Hometown: Cloquet, New Mexico.
College: University of Wisconsin, B.S. in meteorology.
Career Track: Another new recruit, Saeland recently joined The Weather Channel after doing weather in Evansville, Indiana (WEVV-TV), and Hastings, Nebraska (KHAS-TV).
Style: A welcome addition to the testosterone-heavy staff, Saeland is a pro. One of the few meteorologists on the network from west of the Mississippi.

★ Dave Schwartz

Hometown: Philadelphia, Pennsylvania.
College: Temple University, B.S. in meteorology.
Career Track: Dave is one of the more interesting meteorologists on the network. Before getting into television, Dave was a social worker for the Fulton County Health Department. Dave did radio broadcasts for the network for three years before joining the on-air staff at The Weather Channel in 1991.
Fun with Bios: Among other outside interests, Dave lists as his hobbies "singing" and "creating his own greeting cards." He also is an English tutor for Russian immigrants.

★ Lisa Spencer

College: Memphis State University, B.A. in broadcast communications and an M.A. in atmospheric geography.

Career Track: Lisa started out in radio in Memphis, Tennessee, as a morning news anchor and music director. She jumped to TV in 1988 as a weathercaster at WHBQ-TV in Memphis and then on to The Weather Channel in 1991.

★ Mike Seidel

Hometown: Salisbury, Maryland.

College: Salisbury State College, B.S. in mathematics. Mike also has a master's degree from Penn State.

Career Track: Before signing on to The Weather Channel in 1992, Seidel worked at as a meteorologist in Greenville, North Carolina, at WYFF.

★ Dennis Smith

Hometown: Kansas City, Missouri.

College: University of Oklahoma, B.S. in meteorology.

Career Track: Smith was a self-employed "meteorology consultant" to TV stations and corporations before joining The Weather Channel at its launch in 1982. His on-camera experience includes positions at stations in Oklahoma City and Wichita, Kansas.

Style: Dennis Smith is the roving reporter for the network, traveling to the site of severe weather. During one live broadcast, Smith strapped himself to a hotel balcony rail with a belt to keep reporting during a hurricane.

★ Terri Smith

College: Florida State University, B.S. in meteorology.

Career Track: Smith started her TV career as a weathercaster and science reporter at WALF-TV in Albany, Georgia. She did the weather for WXIA in Atlanta before jumping to The Weather Channel in 1991.

★ Marny Stanier

Real name: Mary Leslie Stanier. "My sister always called me Marny when she was little and it just stuck," she explained.

Hometown: Atlanta, Georgia.

College: Stephens College (Missouri), B.A. in communications, 1984.

Career Track: Marny started as a weathercaster in Missouri in Columbia and Joplin before moving to Portsmouth, Virginia, to do weekend weather and reporting for WAVY-TV. The Weather Channel hired her in 1987.

Style: Watch Marny do the weather just once, and you'll realize why she's one of the network's biggest stars.

★ Keith Westerlage

Hometown: La Marque, Texas

College: University of Texas, graduated in 1983.

Career Track: Keith was hired by The Weather Channel in 1984 as a behind-the-scenes graphics person. He left in 1985 to do the weather in Flagstaff, Arizona (KNAZ-TV), and Sioux City, Iowa (KCAU-TV). Keith returned to Atlanta for good in 1987 when The Weather Channel hired him as an on-camera meteorologist.

Style: Maybe I'm biased since he's a fellow native Texan, but I just like Keith—his smooth presentation and slightly goofy smile make him one of my favorites.

Part Two

outsmarting the weather for under $500

5

Weather Stations for Home, Business, and School

I T WAS A DARK AND STORMY NIGHT. As the thunder crashed outside your window, you frantically flipped on The Weather Channel to see what was going on. After you waded through a dozen commercials for Garden Weasels and Salad Shooters, you finally saw a report on your area's "local conditions." Why, there's no storm, and everything's just fine, thank you. You shook your head as the storm raged outside and tossed your patio furniture into the next county.

How could this happen? The "local conditions" reported on The Weather Channel and area weathercasts are from the National Weather Service's local office. The problem is this office is usually located at an airport. But how many people live at the airport?

When former Speaker of the House Tip O'Neil observed that "all politics are local," he could have been talking about the weather as well. Weather just doesn't fit neatly in a box. One side of town may be basking in sunshine while the other half is being drenched by a monster thunderstorm. Or a killer frost might zap your tomato plants because your neighborhood lies in a low spot that is susceptible to cooler temperatures.

Wouldn't it be neat if you could tell exactly what's going on with the weather at *your* house, school, or place of business? In years past, professional meteorological instruments cost several *thousands* of dollars, putting them out of reach of the average hobbyist, teacher, or business person. However, there is good news: today many companies make affordable digital weather stations that are a snap to set up and use. Recent advancements in microchip design have enabled these firms to make accurate *and* affordable weather stations, some of which can hook to a personal computer to make weather watching even more fun and educational.

Using a home weather station has some practical aspects, besides the thrill of seeing a storm send a wind gauge racing as the barometer drops like a rock. Worried that your plants won't survive a severe frost? Many of the stations reviewed in this chapter have an alarm that warns you when the temperature dips to a pre-set level, enabling you to cover the plants or bring them inside *before* it gets too cold. Gardeners will also love the automatic rain gauges, which provide an electronic readout of exactly how much rain you received today, this month, or even this year to date. I use my weather station to warn me of wind gusts over 30 miles per hour so I can retrieve our patio furniture before it blows to Kansas.

One of the first steps to predicting the weather is figuring out what's going on right now, right at *your* home. This chapter reviews stations that not only tell you this but also create a history of your home's weather. Is rainfall below average for this month? On average, how much cooler is your weather compared to the "official" observations at the National Weather Service? You can get answers to these questions and much more by having a home weather station.

Of course, weather stations have more uses than just the home applications described above. In this chapter, I'll review some innovative weather stations for schools, complete with lesson plans and informative computer graphics, designed specifically for educators to use in teaching students about meteorology. This chapter will also take a look at stations for the professional or industrial user, such as a utility company that needs to forecast electricity demand or a ski resort that's praying for snow.

Since each of these uses requires different levels of accuracy and reliability, let's look at them in more detail.

The Three Categories of Weather Stations

Weather stations come in all shapes and flavors. Some offer precision and accuracy, while others trumpet their ease of use and affordable prices. The basic criteria used in this chapter to evaluate a weather station are accuracy, reliability, ease of use, and affordability. I've divided the potential uses of a weather station into three categories:

1. Weather on the Home Front: Weather Stations for Fun & Education. First and foremost, weather stations designed for home use must be affordable. Few people can afford to dump thousands of dollars into weather instruments. Fortunately, many of the new digital stations on the market today can be purchased for under $1000 and some cost less than $500. Moreover, a home weather station should be fun to use and easy to set up. Good display units should convey weather information in a single glance. In my evaluation of such stations, those that combined these attributes and were still relatively reliable and accurate scored high on my list.

2. The 9-to-5 Weather Station: Professional & Industrial Applications. More businesses need access to weather data than you might think. While an airport has an obvious need for weather information, other types of operations are greatly affected by the weather as well. Ice-cream shops can correlate a surge in sales with a blistering heat wave, while ski resorts might have to close a lift if the wind gusts over a certain speed. Many cities have emergency management offices that need up-to-the-minute weather data to warn citizens of floods, severe storms, or other calamities.

The key to a good weather station for such applications is accuracy and reliability. The sensors must be able to withstand harsh operating conditions with little or no maintenance. A display unit is less critical for these uses, since most industrial or professional users will probably hook up the weather station to a computer to help analyze the data. Therefore, stations that have good computer interfaces and software scored higher in my evaluation.

3. The Three R's: Weather Stations for the Educational Market. Many schools have weather "units" built into their curriculum. But how much fun is it learning about the weather from a textbook? A "live" weather station injects a dose of reality into lesson plans. The best weather stations for schools combine educational lessons with snazzy computer graphics to make learning about the weather more fun. In this section, I'll review a weather station that's designed specifically for the education market.

Weather 101: The Basic Functions of a Weather Station

Since even the least expensive weather stations will require an investment of hundreds of dollars, it's important to know what you're buying. Here's a wrap-up of the basic functions and features of the weather stations reviewed in this chapter.

1. TEMPERATURE. Most stations have sensors to monitor both indoor and outdoor temperature. Some even have the ability to monitor a third, or auxiliary, temperature, which can be used for a greenhouse, spa, or whatever. Most outdoor temperature probes have ranges well below zero and over 100° F. If you live in a climate that sees dramatic temperature extremes, check the operating range before you buy.

☺ *Shopper tip:* It's more fun to have a station that reads temperature in tenths of a degree. That way you can dazzle your friends with the knowledge that yesterday's high was 89.7°F, instead of the 90° reading reported at the airport. Watching the temperature toggle up or down is especially interesting during a sharp change in the weather.

☺ *Accuracy Advice:* The outdoor temperature will vary dramatically depending on where you place the sensor. The best place to put the sensor is on the north side of a building in a shady area about four to six feet off the ground. Some of the fancy stations reviewed in this chapter have sensors designed to be mounted on a roof—make sure you don't mount it near (within 30 feet of) an exhaust fan or air-conditioning unit since these obstructions will bias the temperature. Many of the roof-top models have sensor shields that help protect the ther-mometer from direct sunshine. If you mount a temperature sensor on the roof, make sure the station has a calibration fea-ture that lets you fine-tune your temperature readings.

2. HUMIDITY. Many weather stations measure the relative humidity, which is an indication of the amount of moisture in the air. The Earth's atmosphere is like a sponge that can hold only so much water. Humidity gives a reading on how much water is in the sponge. Interestingly enough, measur-ing humidity electronically has been one of the biggest tech-nical challenges for the weather instrument industry. As a result, humidity sensors tend to be expensive, additional "options" for weather stations, ranging from $50 to $500 or more. Some weather stations have the capability of measur-ing both indoor and outdoor humidity.

✪ *Shopper Tip:* If you decide to buy a weather station that mea-sures humidity, make sure it can also calculate the dew point. (Some companies sell separate dew point sensors.) The dew point is the temperature at which the air is completely saturat-ed, or incapable of holding more moisture. You can use the dew point to determine how "juicy" the atmosphere is—a prime ingredient for thunderstorms. A quickly rising dew point may indicate a storm is on the way. When the tempera-ture and dew point are the same, the humidity is 100%.

✪ *Accuracy Advice:* Follow many of the same tips in the temper-ature section above for accurate humidity readings. You need a shady location on the north side of a building. Some humidity sensors must be shielded from rainfall or precipitation.

3. BAROMETRIC PRESSURE. Before the digital revolution, air pressure was measured with mercury barometers. The pres-sure was determined by observing the air's effect on a col-umn of mercury, measured in inches. Many of the weather stations reviewed in this chapter have done away with mer-cury in favor of electronic barometric pressure sensors (usual-ly located inside the display panel).

The key to using a barometer to forecast the weather is to follow the *trend*, rather than using the reading at a specific point in time. Is the pressure rising sharply? This might indi-cate that fair weather is on the way as a high pressure system builds into your area.

Of all the features a weather station offers, barometric pressure is probably the most important tool for predicting the weather. While most weather stations offer a barometer, some of the lower-cost models don't. My recommendation:

don't skimp. A weather station without a barometer is like a pair of binoculars without the lenses—you're not going to see very far.

❂ *Shopper Tip:* Many stations offer barometers with readings in both English and metric units—that is, inches and millibars. This may be helpful if you plan to pull down weather charts from the National Weather Service via your computer (see Chapter 8). The National Weather Service tends to use millibars as a measure of high and low pressure systems.

❂ *Accuracy Advice:* Ever notice that the air in the mountains tends to be "thinner" than air at sea level? This is caused by lower air pressure. As a result, barometers set at sea level must be adjusted for use at high altitudes. Look for instructions on how to calibrate the barometer for your altitude—this may be as simple as calling your local National Weather Service to get the pressure reading. Or you might have to follow some other calibration procedure.

4. WIND. Wind speed and direction are particularly useful pieces of the weather puzzle. A dramatic wind shift might indicate that a cold front has passed your location. Along the Great Lakes, strong northwest or northeast winds may be a harbinger of a heavy lake-effect snow. In Texas and much of the South, a southeast breeze always brings moisture from the Gulf of Mexico—a key ingredient for firing up thunderstorms in the spring and summer. Hence, being able to pinpoint the wind direction and speed will help you forecast future weather trends.

Many weather stations automatically calculate wind chill, which is what the temperature on your skin feels like when you factor in both the wind and temperature. Dangerously low wind chills of -25°F to -100°F can whip across the Great Plains in the winter—knowing what the wind chill is at your location may help with decisions about outdoor activities.

❂ *Shopper Tip:* The technical name for an instrument that measures wind speed is an anemometer. While most anemometers look alike (the wind spins little cups that are attached to a base), the technology used to actually measure the wind can vary greatly from company to company. If you live in an area that is susceptible to windstorms or hurricanes, make sure the upper wind speed range on the anemometer is at least 125 to 200 mph. Your house may not be standing after one of these storms, but your wind gauge will be working perfectly.

Even if you don't get hit by a tropical storm, anemometers installed in coastal areas will suffer from another problem: sea salt will require you to replace the sensors every one to three years. Why? The salt from sea breezes dries up the grease that lubricates the bearings in the sensor—as a result, the anemometer will seize up and stop working. A smart shopper tip if you live near an ocean: make sure to check the availability and cost of replacement parts.

⊙ *Accuracy Advice:* The wind speed and direction sensors should be mounted on a pole about ten feet from the top of your roof. If you use a lower location, you might get false readings from the wind bouncing off your roof (the technical term for this is refraction). Obviously, it's best to make sure the area is not obstructed by trees, chimneys, or any other object. In reality, installing a weather station at a home involves some trade-offs. If your home stands in a grove of tall trees, your wind readings will be lower than those in an open area. Nonetheless, a wind sensor will still provide valuable information.

If you mount any weather sensors on your roof, you've instantly created a lightning rod—be sure to ground the pole and sensors with appropriate wires (visit a local electronics store like Radio Shack for the supplies). And speaking of lightning, you may want to make sure your weather station (and modem) have a surge protector to protect against damaging electrical spikes that travel down electrical and phone lines.

5. PRECIPITATION. Measuring the exact amount of rainfall is one of the best benefits of a weather station. Rain can be deceiving—a quick downpour may look impressive but only drop a minuscule amount of rainfall. Conversely, a long, slow rain for several hours may dump significantly more moisture. Many weather stations offer rain collectors as an affordable option—usually under $100.

⊙ *Shopper Tip:* The best rain gauges are self-emptying—rain is collected in a bucket, counted in a sensor, and then released out a hole in the bottom of the gauge. All of this is done automatically, without any moving parts or batteries to replace. I also like rain collectors that are precise to a hundredth of an inch (.01"), the standard for measuring rainfall. You should be aware that many rain collectors only work when temperatures are above freezing—some manufacturers recommend you cover the collector during the winter. Why? A few rain gauges have glass components that are susceptible to breakage if water in the gauge freezes. You might want to check this detail before purchasing a rain gauge if you live in a cold climate.

⊙ *Accuracy Advice:* The placement of a rain sensor is a tricky matter. Mounting one on a pole on the roof often isn't a good idea: air currents at that level have been known to interfere with rain collection, and some rain collectors (most notably, Rainwise's rain gauge reviewed on page 86) can give false high rain readings caused by the pole vibrating in a storm. You also need access to the rain collector from time to time for maintenance (to clean out debris, etc.). A better bet: mount the rain collector two or three feet from the ground on a level surface. Make sure that the rain can drain out the bottom of the collector.

In reality, you may have to make some compromises. Schools should probably mount a rain gauge on the roof to avoid vandalism. And other rain collectors are less sensitive to pole mounting on a roof—I've had a Davis rain collector (see review later in this chapter) mounted on a roof for a year and have had relatively accurate readings.

6. OTHER FUNCTIONS. The entrepreneurial zeal of many weather station manufacturers has given us a plethora of functions measuring more obscure aspects of the weather. One of the more recent hip instruments is a light sensor. In our review of the Automated Weather Source later in this chapter, you'll notice this station has a light sensor that produces a "sunshine index" (the amount of sunlight hitting the Earth). Here's a wrap-up of some other common features:

• Go Metric. Most stations let you show temperature in Celsius degrees or the barometer in millibars.

• Reach New Highs (And Lows). You can find out today's high temperature or yesterday's high wind gust, all with a touch of a button. Best of all, many stations will tell you the exact time and date of each high or low.

• Get Alarmed. Gardeners will love this feature—pre-set the weather station to sound an alarm when the temperature (or any other function) falls to a certain level. That way you can be reminded to retrieve your geraniums when a sudden storm sends the temperature plunging.

• Do the Cyberspace Bop. Most weather stations can "talk" to personal computers. This feature has added a new dimension of understanding to the weather—your computer can store information, enabling you to graph and analyze data in ways that show weather patterns in a new light. For example, you can graph a cold frontal passage, tracking how the barometer starts to rise at the exact time the temperature falls. The best weather stations have affordable software programs (for either IBM or Macintosh computers) that are easy to use.

How to Order a Weather Station

Most of the weather stations in this chapter can be ordered directly from their manufacturers. Or, if you prefer, you can order one through the many weather catalogs listed in Chapter 11. The advantage to catalog shopping: some offer discounts from the retail price that can save you hundreds of dollars. If a catalog does offer a discount for a weather station, this fact is noted in the box at the end of each station's review.

Before investing any money, call the company and ask them to send you literature on the station. Most have brochures with detailed information on accuracy, the station's construction, and so on. Given your location's climate and the particular application you're considering (i.e., home, professional, school), you should ask the company questions about the unit's reliability, warranty, etc.

The Ratings

★★★★ EXCELLENT—*my top pick!*

★★★ GOOD—*above average quality.*

★★ FAIR—*could stand some improvement.*

★ POOR—*yuck! could stand some major improvement.*

WEATHER ON THE HOME FRONT:
Digital Stations for Fun & Education

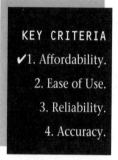

KEY CRITERIA

✔1. Affordability.

2. Ease of Use.

3. Reliability.

4. Accuracy.

★★★★ *The Best*
Davis Weather Monitor II

My pick for the best home weather station is the Davis Weather Monitor II. This affordable weather station performs an amazing number of functions for a price that's quite affordable.

The basic model ($395) measures inside and outside temperature (-50°F to 140°F), wind direction and speed (to 175 mph), wind chill, inside humidity, and barometric pressure. Add outside humidity/dew point ($125) and a rain collector ($75) to make a complete weather station. The total cost with all the options: $595.

Davis is a leader in making weather stations more affordable and fun to use. The Weather Monitor II's display unit is a case in point. Wind direction and speed are shown in an easy-to-read directional compass via a flashing arrow and a large readout, respectively. In addition to these features, you can display one other function (temperature, barometric pressure, or whatever you find most interesting). There's even a trend arrow to indicate whether the barometric pressure is falling, rising, or steady.

And that's only the beginning. Did you want to know what was the highest temperature today? Press "recall" and the Weather Monitor II will tell you the high (or low) temperature, plus the exact time it occurred. You can also get maximum and minimum extremes for humidity, dew point, and wind chill. Prefer metric units to English? The Weather Monitor II can display every function in metric, including wind speed (meters per second or knots), barometric pressure (millibars), and rainfall (millimeters). The display unit also has a scan that enables you to automatically cycle through several readings during a short period of time.

The unit's alarm function is impressive. If you're worried that your tomato plants might freeze during a cold snap, you can have the weather station remind you when the temperature falls to a certain reading. In addition, you can set alarms for wind speed, wind chill, and humidity.

Well, Nobody's Perfect

The Weather Monitor II isn't perfect, and after using it for over a year, I have discovered some bothersome faults. First, the display unit doesn't like static electricity. Touch the unit on a dry winter's day and zap! Static electricity will zap out the unit's display. You'll have to reset the entire unit (including disconnecting the battery backup) and reenter the date, time, and barometric pressure.

A few other problems: the Weather Monitor II doesn't automatically reset the highs and lows at a certain point each day, so you must manually clear the memory every day. If you forget, you may be looking at a high or low reading from a couple of days ago. An optional automatic reset would make the station much easier to use. In addition, unlike Maximum's WeatherMAX (see review later in this chapter), the screen doesn't swivel. Also, the light on the display is somewhat difficult to turn on (you have to depress any two buttons at once). Occasionally, I've found the light turned on all by itself, which poses a problem since the display light should not be left on for a long period of time or it will distort the inside temperature reading.

A couple of the Weather Monitor's other flaws have been

fixed by Davis. A lousy rain collector that could only read rainfall down to one-tenth of an inch has been replaced by a new one that has an accuracy of .01"—the standard for measuring rainfall. Also, Davis has replaced its finicky humidity sensor with a more accurate and reliable model. I bought one of those old humidity sensors for $125 and wasn't pleased that it fizzled out after just a couple of months of use. I wish the company would have notified me by mail of the defective sensor—at these prices, we should at least have the option of upgrading to a new sensor at a low price (or better yet, receive a replacement sensor free).

Customer service at Davis is good but could stand improvement in a few areas. On the upside, the company does occasionally offer existing customers "upgrade" discounts to trade up to new equipment. On the downside, I think their shipping charges are exorbitant—they are based on the dollar value of the order instead of its weight (a mistake in my opinion). Order a $6 replacement part, and you'll be slapped with a $5 shipping charge.

Accuracy and Reliability: Do You Get What You Pay For?

How reliable and accurate can a $400 home weather station be, especially when competitors' products cost two and three times that amount? This question seems to be open for debate among the weather buffs I interviewed for this book.

On one hand, let's be realistic. You can't expect the accuracy of professional instruments (used by the National Weather Service) from low-cost home weather stations. Professional instruments costs thousands of dollars, and one would expect them to deliver more. On the other hand, just because a home weather station doesn't measure barometric pressure as accurately as expensive barometers doesn't mean it isn't useful. The key to using pressure readings to monitor the weather is the *trend* of the reading, not so much the actual figure. If the barometer is off by a tenth of an inch, it doesn't affect your ability to forecast the onset of a major storm.

All of this theoretical discussion aside, I actually field-tested a Davis Weather Monitor II for a year in some pretty tough weather—hurricane-force windstorms, winter temperatures that plunged below zero, one-foot blizzards, summer thunderstorms with hail and strong winds. . . yep, just a typical year in Boulder, Colorado. How did the Weather Monitor II hold up? Pretty good, as it turns out. Oh sure, during one windstorm with 60 mph gusts, the top of the rain collector blew off to parts unknown. After that incident, I decided to glue the rain collector to its base (though this impedes the suggested annual cleaning and maintenance). All of the other sensors seemed quite durable and very accurate, despite the station's overall low cost.

Davis' Other Products:
The Perception II and Weather Wizard III

If the Weather Monitor II has more bells and whistles than you need, Davis also sells two other models with stripped-down features. For example, the Perception II is designed to "monitor indoor climates." Temperature, barometric pressure, and humidity are displayed simultaneously. There's no installation—just plug it in, set the time, and adjust the barometric pressure to your altitude. Like all of Davis' products, the unit records the highs and lows of most functions and has alarms for temperature, humidity, and time. A clock, date, and back-lit display feature rounds out the Perception II's offerings.

Perhaps the most valuable feature of the Perception II is the station's barometric trend arrow—the display instantly tells you whether the pressure is rising, falling, or steady. Such information is crucial to the forecasting of coming storms. At $150, the Perception II isn't the best bargain available, and other indoor weather stations like the Pacesetter Six Weather Station do more for less. (In Chapter 11 there is more information on the Pacesetter, which is available from several catalogs. See page 169).

The Weather Wizard III is a scaled-down version of Davis' Weather Monitor II. The Wizard features temperature (inside and outside), wind speed and direction, and wind chill. At $195, this isn't a bad deal when you consider a wind speed and direction sensor alone can cost that much or more.

Options for the Wizard include a rain collector ($75), which lets you monitor daily and accumulated rainfall, and computer storage, analysis, and graphing of your home weather data using Weatherlink ($165; see below for more information). My biggest gripe with the Wizard: no barometer or humidity sensor. Without a way to monitor barometric pressure, your ability to forecast weather trends will be severely limited. The lack of a humidity sensor (and, therefore, dew point readings) may also disappoint some users. A rising dew point and sharply falling barometer might be the harbinger of a major storm—hence, while the Wizard's temperature and wind functions are helpful, they only give half the story.

Computer Compatibility

One of the most powerful features of Davis' weather stations is their computer compatibility. You can connect any of their stations to a personal computer with Davis'

Weatherlink program. Best of all, Weatherlink comes in *both* IBM-PC and Macintosh versions. Weatherlink includes a data interface that connects the weather station's display unit to your computer and the software to make it all work. You can create graphs, analyze trends, and plot data.

The program creates a current weather "bulletin" that displays all current weather functions—temperature (inside and out), dew point, wind speed and direction, humidity, barometric pressure, and rainfall (daily and monthly totals). The bulletin screen is useful when a major weather event is occurring, such as the passage of a cold front or thunderstorm. You can watch the wind speed increase, the temperature fall, and the barometer rise—all at once. In fact, the bulletin screen displays the barometric trend on a graph over the past four hours, quite useful for tracking major storms.

Weatherlink stores your home weather data on your hard drive. You can create a historical log of your data, graph one month's highs and lows, determine if rainfall is above average, and more. The Weatherlink program stores data at an interval you select, as frequent as every minute.

After using Weatherlink for a year, I have found it to be an amazing tool. The best part: you can graph two functions at once (e.g., barometric pressure versus temperature) to see how different aspects of the weather interrelate. Unfortunately, it does have its shortfalls. The graphics (especially in the bulletin screen) could be much better—a thermometer graphic would be better for the temperature reading than Weatherlink's boring, two-dimensional bar graph. I'd also like to adjust the barometer graph to show readings from more than just the last four hours. Prevailing wind direction and average wind speed information would also be good additions.

As for the data log, I would like to be able to add snowfall totals to my own station's log. Unfortunately, you can't edit the data log or add to it (though you can delete individual time records). A new version of Weatherlink for IBM-PC or compatible computers allows you to edit the data, but the Macintosh program still lacks this feature. Weatherlink could also be improved by adding three-dimensional graphics as well.

Despite these minor drawbacks, Weatherlink is a great addition to your weather-predicting arsenal. In Part II of this book, I'll explain how to use Weatherlink to outwit storms and predict weather changes.

The Bottom Line

The Weather Monitor II is one cool weather station. For about $600, you can monitor the entire spectrum of weather at your house, including temperature, rainfall, wind, pressure, and humidity. Considering that other stations with similar capabilities are well over $1000, Davis offers an excellent value for the weather junkie.

Got a computer? Davis' affordable software, Weatherlink, will give you the ability to store, analyze, and (most fun) graph your home weather station's data. I like the ability to graph two functions at once (comparing, say wind direction and temperature), although the graphics themselves could be a little more snazzy.

If your budget doesn't allow for the Weather Monitor II and all you want to watch is temperature and wind, the Weather Wizard might be a good station to consider—at under $200, it's a third the cost of Davis' more expensive station. The Wizard also allows you to add a rain collector as an option. On the downside, the lack of a barometer and humidity sensor will frustrate forecasting.

Davis leads the market in offering affordable yet accurate home weather stations. I highly recommend their products.

Information Box
WEATHER MONITOR II
Rating: ★★★★
List Price: Weather Monitor II $395. Perception II $150 Weather Wizard III $195.
Options: Rain collector $75. Humidity sensor $125. Weatherlink software $165.
Company: Davis Instruments, 3465 Diablo Ave., Hayward, CA 94545. (800) 678-3669 or (510) 732-7814. Fax: (510) 670-0589.
What's Cool. Low cost, fully functional weather station for under $600. Excellent graphing and storage software has both IBM and Macintosh-based versions.
What's Not. Some sensors in the past have lagged in accuracy and reliability. High shipping charges for replacement parts. Display can be zapped by static electricity.

★★★ *Runner-up #1*
WeatherMAX

When I heard that Maximum was planning to debut WeatherMAX, an affordable computerized weather station, I was skeptical. Maximum is famous for its gorgeous (read: expensive) brass weather instruments. For example, the Maestro, a simple wind speed and direction instrument weighs in at *over* $400. And that's just to measure the wind.

So, I was pleasantly surprised to see Maximum bring in the WeatherMAX at under $500. Did they skimp on the quality of the sensors? No, the company claims to use the same "high-quality sensors (we) use with our brass instruments, including the Mt. Washington-tested anemometer."

Basically, you get a package that is very similar to the

Davis Weather Monitor II. The WeatherMAX features wind speed and direction, indoor and outdoor temperature, wind chill, barometric pressure, indoor relative humidity, time/date, and alarms for various functions (if the wind exceeds a certain speed, for example). Besides current readings, the WeatherMAX gives you averages, highs, and lows for most of the functions.

So what does all this cost? The WeatherMAX runs $495. Options include an outdoor relative humidity sensor for $195. In addition to humidity, this sensor also measures the outdoor dew point and records highs and lows. Another optional accessory is a rain gauge for $100. Basically, if you go for the whole enchilada, you'll be out $790. Since Davis sells a similar system for about $600, you may wonder, "What does Maximum do for this extra $200?"

First, I was impressed that the WeatherMAX rain gauge has a resolution of .01"—Davis only recently introduced a rain gauge with similar precision. The WeatherMAX also offers three different memories for rainfall (daily, weekly, and yearly). Davis only offers current and total-to-date readings. If you need different time periods than the standard daily, weekly, or yearly settings, the WeatherMAX lets you choose any three time periods you desire—a nice plus.

I also like that the WeatherMAX will store high and low readings for as long as you like—or the station will automatically reset all the memories every day at the time you specify. The Davis Weather Monitor does not offer this "automatic reset" feature, and trust me, having to reset the unit every day becomes a hassle quickly.

The WeatherMAX is not perfect. I wasn't impressed that the outdoor humidity sensor (which costs over 50% more than Davis' similar unit) doesn't measure humidity below 20%. If you live in an arid climate where the humidity frequently falls to the single digits, you may want to think twice about WeatherMAX. Similarly, the dew point doesn't measure figures below 0°F—once again a bummer for those in dry climates.

The Better Display? Advantage: Davis

It's obvious that the engineers at Maximum must have been "inspired" by the Davis Weather Monitor II's display. The WeatherMAX is a carbon copy, down to the wind speed displayed inside the wind direction compass. Like Davis, you can only look at one other function at a time besides wind.

I have mixed feelings about the trend arrows for the barometer and temperature. I like the arrow that indicates whether the temperature has been falling or rising—a feature Davis doesn't offer. Yet, the trend arrows themselves are tiny. In fact, much of the display features small readouts or numbers that may be hard to decipher at a distance.

At least the unit is compact—at six by seven inches, you can use the WeatherMAX on a desk or mount it on a wall. One neat feature: the display screen swivels 180° for easier reading.

Computer Link? You're on Your Own

Another major bummer for the WeatherMAX is its lack of computer compatibility. While Maximum includes a serial port (RS232) jack, there is no software available to log or graph data at the time of this writing. While the company offers "suggestions for developing your own software interface," for $800 they should figure out a way to do this for us. Unless you're proficient at building a database from scratch, you may want to skip the WeatherMAX if you have your heart set on using your weather station with a computer.

As this book went to press, I heard that Maximum is planning to release software for IBM or compatible PCs for the station to log and graph data. The company said they were debugging the software and hoped to have it on the market in late 1994.

Maximum's Brass Instruments

If you're the CEO of a Fortune 500 company and want to hang some expensive-looking weather instruments on your office wall, you'll be impressed with Maximum's fancy brass instruments.

The company offers 15 different instruments that measure everything from barometric pressure and tides to wind speed and humidity. You can buy single instruments ($165 to $595 each) or group them in a wood-mounted weather station (oak or mahogany finishes are available). I was impressed with the sheer variety of different options available—you can mix and match the instruments to your heart's content.

For example, the "Professional" is a five-instrument panel with wind speed/direction, barometric pressure, outdoor temperature, indoor temperature/humidity, and time. This unit will set you back about $1600. I should note that

many of the catalogs that I'll review in Part IV of this book sell the Maximum brass instrument stations at a significant discount. For example, the Northeast Discount Weather Catalog (800-325-0360) sells the "Professional" station for about $1200.

The Bottom Line

The WeatherMAX is a good second choice for a home weather station. Even though it costs $200 more than a similarly equipped Davis Weather Monitor II, Maximum's reputation for reliability and accuracy may be worth the extra price if you live in a harsh climate. The rainfall collector, for example, has a resolution of one one-hundredth of an inch.

While those in rainy locales might love the WeatherMAX, weather buffs in dry climates may curse it. The expensive humidity sensor's failure to measure humidity below 20% will be a disappointment. Since extremely low humidity in the West often signals a high danger of fire, this shortcoming may be a big headache for volunteer fire fighters looking to a weather station for climate data.

If you want to use your computer to log historical data, Davis' WeatherLink software puts their station head and shoulders above Maximum's WeatherMAX. While the WeatherMAX unit does have a built-in serial port, the lack of accompanying software is a major negative.

Despite these quirks, the WeatherMAX still scores relatively high on our "fun meter." The trend arrows for barometric pressure and temperature plus the automatic reset feature will make weather-watching an enjoyable experience.

Information Box
MAXIMUM WEATHERMAX
Rating: ★★★
List Price: WeatherMAX $495.
Options: Outdoor relative humidity sensor $195.
Rain collector $100.
Company: Maximum, 30 Samuel Barnet Blvd.,
New Bedford, MA 02745. (508) 995-2200.
Best Price in a Catalog: The Northeast Discount Weather Catalog (800-325-0360) sells the WeatherMAX for $399, the humidity sensor for $165, and the rain collector for $85. The total for the entire set-up would be $650—not bad. The Weather Station catalog (603-526-6399) sells the outdoor humidity sensor for just $150.

What's Cool. Lots of neat features like prevailing wind direction and a temperature trend arrow. Display swivels for easy reading from any angle. Has automatic reset option that clears out highs and lows at a time you specify.

What's Not. About $200 more than the competition for similar features. Lousy humidity sensor. Display readings could be more readable. No software for logging or graphing data.

★★★ *Runner-up #2*
Weather Report

Texas Weather Instruments makes the Weather Report, which is a nice weather station except for the fact that it can only be used in the Lone Star State.

Just kidding. Seriously, the Weather Report solves one of the drawbacks to many home weather stations: puny displays. Often, you can view only one or two functions (say, temperature and wind speed/direction) at the same time. Texas Weather Instruments' Weather Report overcomes this problem with an attractive display that shows several functions at once—albeit at a price that may make some weather junkies pause.

The Starter System includes wind direction and speed, barometric pressure, and inside and outside temperature. We found this system to be rather pricey at $1000, considering competitors sell similar stations for as low as $400.

More impressive is the Standard System, which includes all of the above and adds outside humidity and a rain collector for a total of $1279. If you want to go whole hog, the most expensive model ($1699) includes a solar radiation sensor, which detects the amount of sunlight.

An Attractive (and Sizable) Display

Perhaps the best aspect of the Weather Report is its wood-trimmed display unit. Set against a black panel, the station features large LCD readings for time/date, temperature, barometric pressure/humidity, and rainfall. The wind speed indicator is inside a compass that also displays the wind direction—a nice feature that makes monitoring the wind more fun. The only downside to all this: the display unit itself is over a foot wide, six inches tall, and nearly four inches deep—not exactly compact.

A series of small buttons lets you display minimum and maximum readings for a host of functions (including the date and time of occurrence), as well as a button that changes readings from English to metric.

The Weather Report has some unique features. I like the

"comfort index," which shows the temperature/humidity index—a nice feature if you live in a steamy climate where the combination of temperature and humidity may make it seem much hotter than the air temperature suggests. The station also has a tide clock that predicts high and low tides and a function that gives you the last date and time of the most recent rainfall. In addition, the Weather Report has lights that indicate a rising or falling trend in temperature, barometric pressure, and humidity— another big plus for forecasting.

Computer Link

The Weather Report is equipped with a serial port (also known as a RS-232 adapter). You can connect the display unit directly to a printer and program the station to print weather information at various intervals.

Texas Weather Instruments offers two software packages for linking the Weather Report to an IBM-compatible computer. The "Full Report" ($129) graphs, logs, and prints data, and the "Cloud9" ($249) creates graphics, prints data, and lets you access the station from a remote location via a modem. There are even software packages available if you'd like to turn the Weather Report into a voice-based weather reporting system (others can call your station and get weather information over the phone).

I tried out demos of the Full Report and Cloud9 and was impressed. The Full Report program has a current conditions screen with four trend boxes that graph conditions over the past 24 hours. Cloud9 is even more powerful—one screen features six "dials" that display current conditions. Underneath the dials are large red readings of actual conditions (you can customize what's displayed and how). The effect is space age, if not somewhat overwhelming. Cloud9 also includes an extensive graphing program that enables the user to display historical information in either line or bar graphs.

Accuracy

The manufacturer claims that the Weather Report's superior technology lifts the station above competitors. Infrared digital technology enables wind readings to have as little as a 2% margin of error, and there are no posts or reed switches (common in other wind sensors) to wear out. You can calibrate most functions, which helps ensure accurate readings. Texas Weather Instruments also trumpets their humidity sensor, pointing out that their technology is dependable with only a 3% margin of error.

82

Bottom Line: In-Depth Weather for a Not-so-Cheap Price

The Weather Report is a weather station for serious weather junkies with a correspondingly serious bank account. Several features will be helpful for gardeners—an auxiliary temperature probe ($39) can be used to monitor soil temperature or conditions in a greenhouse. The ability to monitor three temperatures at once (inside, outside, auxiliary) is a big plus. Weather junkies who want tidal information will like this unit as well. The computer programs are powerful—Cloud9, with its graphing features, is probably the better bet.

Information Box
WEATHER REPORT
Rating: ★★★
List Price: Starter System (WR-25-C) $999, Standard System (WR-25) $1279, Solar Radiation (WR-25-S) $1699.
Company: Texas Weather Instruments, Inc., 5942 Abrams Rd. #113, Dallas, TX 75231. (800) 284-0245 or (214) 368-7116.
Best Price in a Catalog: American Weather Enterprises (215-565-1232) sells the Weather Report Standard System (WR-25) for $1149.
What's cool. Great display. Unique features like temperature/humidity index and tide clock. Decent software options to graph and display data.
What's not. Expensive. Software has no Macintosh version.

★★★ *Runner-up #3*
4th Dimension History Logging Station

Computer nerds will love the 4th Dimension system, a weather station that was designed with cyberspace in mind.

The system includes a complete sensor package from Rainwise (for more information on this company, see review below), an IBM-PC compatible computer with keyboard and high-resolution monitor, and the 4th Dimension software. Hence, setting up the station also involves also setting up the computer and software—with all the attendant DOS-related brain damage caused by files with names like "AUTO-EXEC.BAT."

The Display

The most exciting feature of 4th Dimension is their

graphical display of weather information. Basically, your computer screen is divided into three sections: current weather, a chart of measurements for the last 24 hours, and a log of long-term weather history. This system's unique way of correlating current and past conditions is intriguing. The current weather information is displayed digitally at the far right side of the high-resolution monitor. At the top is the barometric pressure, which includes a trend arrow indicating any change more than .01" in the past hour. Below the barometer is a temperature display, with a trend arrow and wind chill reading. The wind direction is displayed both on a compass and with letters (such as WSW). In addition to cur-

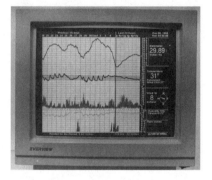

rent wind speed, the software also calculates the mean (or average) wind speed over the past 30 minutes. That's pretty neat. The current weather information displayed also includes humidity, dew point, and rainfall for the last 24 hours (measured at .01").

Another convenient feature of the 4th Dimension weather station is its chart of recent measurements and the log of long-term weather history. Basically, the software creates an electronic strip chart (which is a horizontal chart, like an EKG printout) that plots all measurements as they change over time. The data is plotted five times per hour (though you can adjust this figure to as often as one sample per minute!). Watch the barometer rise after a front passes and notice that the temperature and humidity drop.

Most interesting is the way 4th Dimension treats wind and rainfall. The average wind speed is plotted as well as the minimum and maximum wind gusts. Rainfall totals appear to be "growing" over time—this somewhat complex graph indicates the rate at which the rain was falling.

The historical log "spills" over the 24-hour chart of recent measurements. You can view data from the last two weeks up to a full year at a time on the screen. As a result, you can make overall judgments about the weather, such as whether the last month was cooler or wetter than normal.

All in all, the station seems easy to use, despite the somewhat ponderous instruction manual, filled with such arcane information as the screen pixel size. The advice about the weather is also needlessly abstruse—take a look at this quote: "The temporal juxtaposition of changes in the different weather parameters reveals the relationship between pressure and wind, between moisture and rain and the manifold events that characterize the passage of a front." Whoa! Silly me, before I read this I thought "temporal juxtaposition" was a fondue recipe.

A "Lite" Version, too

Let's say you don't want to spend $1795 on the entire weather station. Well, there is good news: Weather Dimensions sells their software separately. It is fully compatible with several other stations, including Davis' Weather Monitor II and WeatherMAX by Maximum (see reviews of those products in this chapter). The software comes in "professional" ($395) and "lite" ($79.95) versions. The lite version does everything the professional program does except it lacks the ability to print current observations and to change the time scale to other than a 24-hour/15-day display. Also, the less expensive version cannot scroll back over the station's accumulated history (what a bummer) or write spreadsheet data beyond the visible 15-day window.

Another letdown: the software only works with IBM-compatible PC's that run DOS (you need a 286 or later computer with a VGA monitor)—there's no Macintosh or Windows version. Also, you must "dedicate" a computer to work full-time logging your weather data—unlike the Davis Weather Monitor II, the system doesn't have a "buffer" that enables you to store data and then download to your computer at your convenience. A spokesperson for Weather Dimensions said they were working on a future version of their software that would incorporate a data buffer.

Despite these shortcomings, the software is still worth a look. If you've got a PC to dedicate to logging weather data, you can use a lower-cost station and Weather Dimension's software for a combined price of less than $1000. Another interesting note: a demo version of the software is available if you want to give it a test drive.

The Bottom Line

While the interface may look somewhat ponderous at first, the 4th Dimension History Logging Weather Station packs a powerful punch. The options for data display seem endless. You can shade (in color, of course) the temperatures for the past month that exceed normal. Or track the number of hours your weather station records readings under freezing.

Who would most enjoy this system? Probably weather buffs who are interested in creating detailed weather logs for their area. Another possible use: businesses that depend on the weather (farmers or ski resorts, for example) that need to monitor weather trends exactly.

What about the educators? While it's possible the 4th Dimension History Logging Weather Station may appeal to some teachers, its lack of educational materials and dense instruction manual will be a major negative.

Information Box
FOURTH DIMENSION HISTORY LOGGING WEATHER STATION
Rating: ★★★
List Price: $1795, includes the sensors, processor, and color display.
Company: Weather Dimensions, Inc., PO Box 846, Hot Springs, VA 24445. (800) 354-1117.
What's Cool. Strip-chart history of weather data. Unique features include average wind speed and rainfall rates.
What's Not. No Macintosh version. Somewhat ponderous line graphs and tiny readings (the wind direction arrows are microscopic, for example) may turn off some users. Instruction manual could be more user-friendly. High cost.

★ *Runner-up #4*
Rainwise

I'm sure this weather station was pretty cool when it debuted several years ago. Unfortunately, its clunky display is now as hip as an orange '75 Plymouth hatchback with a rusting hood.

The sad thing is that the weather station itself isn't half bad. The innovative sensor unit features a combination self-emptying rain gauge and wind speed sensor. The wind direction finder resembles a space ship with its sleek black housing.

Everything You Wanted to Know About the Weather (and a couple of things you don't)

The WeatherStation has a decent selection of features, including indoor/outdoor temperature, barometric pressure, wind direction, wind speed, wind chill, outdoor relative humidity, and rainfall. There are also a couple of worthless readings—the unit calculates "degree days heating and cooling," for example. Cooling degree days is defined as the figure that "any time the average temperature for a 24-hour period is more than 65°F, (this) additional amount is stored in memory." The result is an indication of how much you have to use your air conditioner. So what? Why Rainwise opted to include heating and cooling degree days and leave off inside humidity is beyond me. On the upside, the rain gauge does measure rainfall with a resolution of one one-hundredth of an inch.

Clunky Display

All this great weather information is wasted on a display

panel that has serious limitations. The only display gauge on the large and narrow panel (7" wide, 17" long by about 2" deep) is a big, red LCD screen in the unit's upper right corner. Hence, you can only see one weather function at a time—major bummer. For example, to go from the outside temperature to the wind speed, you have to turn a large silver knob. Since there is no display compass, wind direction is displayed in degrees, which is cumbersome. How much fun is it to try to remember 220° means winds are from the south-southwest?

The Digital WeatherStation also has no arrows or lights to indicate a rising or falling barometer. Instead the whole display blinks if a rapid fall in pressure is detected. If the pressure is rising rapidly, just the first two digits blink. Yuck.

Other Products Fill Niches

While I think the WeatherStation is a loser, Rainwise scores better with WeatherVideo. This product lets you view weather information on any television. A small video processor sits next to your TV and connects with Rainwise's rooftop sensor unit. The result is an interesting TV display of wind speed, direction, temperature, wind chill, barometric pressure, humidity, and rainfall. WeatherVideo costs $990, which includes the sensors listed above and the TV interface.

Despite the usability of the system, Rainwise WeatherVideo is hobbled again by poor engineering. For example, in order to display maximum or minimum temperatures, you're supposed to touch the screen with a light pen that's attached to the video processor. A remote control would have been a better choice here—when does anyone walk over to their TV anymore?

And the display's graphics aren't anything to brag about. The biggest graphic element is Rainwise's corporate logo (an owl) inside the wind direction compass—c'mon guys, for this price, you could de-emphasize the corporate advertising. The actual wind speed is a small display inside the logo. Given the small size of the display, the readings are hard to see at a distance.

If you're looking for a more traditional and easy-to-use weather station, Rainwise also offers the Oracle series. Four models are available that include combinations of outside or inside temperature, barometer, wind direction and speed, wind chill, humidity, and other functions. Each variation is

designed for different applications. For example, Oracle 3 is for greenhouses or gardeners and displays outside air temperature, soil temperature, rainfall, growing days, and soil moisture. The display is attractive: Rainwise recesses the black panel in a block of wood. Switching among the various functions involves moving a magnetic button on top of the unit—a novel idea. There is also an "auto-select" feature that scans all of the functions. Unfortunately, the Oracle series' prices are rather high ($439 to $595) for a unit that essentially can only display one function at a time.

Weak Computer Ties

If you want to use your computer to log and graph weather data, you probably should pass on the Rainwise weather products. Only the WeatherStation has a serial port (RS-232C) output option to connect to a computer or printer. And this option adds about $300 to the price of the station.

Unfortunately, Rainwise doesn't offer any software to help you graph and display weather data on your computer—you're on your own.

The Bottom Line

I can only recommend Rainwise's weather stations for particular uses. For example, the WeatherVideo can give couch potatoes a quick look at current conditions. However, Rainwise's flagship product, the WeatherStation, is just too expensive and limited in its use to be of any value to hobbyists.

Information Box
RAINWISE WEATHERSTATION
Rating: ★ for the WeatherStation,
★1/2 for WeatherVideo.
List Price: Digital WeatherStation $1029, Oracle weather stations $439-$595, WeatherVideo $990.
Company: Rainwise Inc., 25 Federal St., Bar Harbor, ME 04609. (207) 288-5169.
Best Price in a Catalog: $990 for the Digital WeatherStation in the Weather Station catalog (603-526-6399).

Honorable Mention: Ultimeter II

This is perhaps the least expensive digital weather station. At just $179, the Ultimeter II offers an impressive range of features: wind speed/direction, wind chill, outside temperature, and more. With an optional rain gauge ($60), you can even monitor daily and monthly rainfall. The Ultimeter II records highs and lows for wind speed, wind chill, and outside temperature—complete with the time and date of occurrence. The best feature: the wind sensors are designed to fit on the top of a TV antenna mast. What's missing: no barometer or humidity sensor. The company even offers computer software and a RS-232 interface ($49 total) to connect to an IBM or compatible PC. Available from the Weather Station Catalog (603-526-6399).

THE 9-TO-5 WEATHER STATION INDUSTRIAL & PROFESSIONAL APPLICATIONS

KEY CRITERIA

✔ 1. Accuracy.
2. Reliability.
3. Ease of Use.
4. Affordability.

★★★ *The Best*
MesoTech Automatic Weather Stations

The most innovative technology award for weather stations goes to MesoTech. Take a quick look at this California-based company's sensor unit.

Amazingly, this unit measures wind speed, direction, air temperature, barometric pressure, relative humidity, dew point temperature, and rainfall. You'll notice that the unit has *no moving parts!* How, then, does it measure wind?

Funny you should ask. MesoTech uses its own patented technique for determining wind speed and direction, called "Thermal Field Variation," which measures the thermal field changes caused by wind passing a heated cylinder. Basically, this works the same as when you hold a wet finger in the air to determine wind direction.

Since the station has no moving parts, it may be ideal for applications in harsh environments. The

Pentagon used a military version of MesoTech's station on tanks during the Persian Gulf War. Those stations cost $30,000 *each*. Fortunately, MesoTech's commercial version costs less—about $2800 for a device the measures the wind and $4000 for one that does everything.

All of the components for measuring the weather are enclosed in a sealed aluminum cylinder. The company even markets a stainless steel version for "marine applications." This steel station, which has "extra protection from salt corrosion and submersions," is suitable for installation on ocean platforms and vessels, according to company literature.

The flexibility of the station is amazing. The temperature probe, for example, can measure temperatures from -40°F to 130°F.

Computer Capability

While I was impressed with MesoTech's high-tech weather station, I have one problem with the display unit. There isn't one. Instead, the sensors have a serial port (RS232 or RS422) that connects to any IBM PC-compatible computer. For this price, you'd expect at least some type of display unit that doesn't require a computer connection. Mike Lydon, president of Mesotech, said that many of their customers use an inexpensive PC-compatible computer to display the weather information. The station includes software for both DOS and Windows applications that allows the user to graphically display weather information. You can also link the data to spreadsheet or database programs for further analysis.

The Bottom Line

MesoTech's weather station is a marvel of high technology. Its breakthrough design with no moving parts makes it perfect for harsh climates or environments. If you have an application where access to the sensor is difficult (such as a high tower at an airport), this unit might be the answer since it requires no calibration or maintenance.

Information Box

MESOTECH'S AUTOMATIC WEATHER STATIONS

Rating: ★★★

List Price: $2800 to $4000.

Company: MesoTech, 4670 Chancery Way, Sacramento, CA 95864. (800) MESOTECH or (916) 483-0600.

What's cool. Look Ma! No moving parts! High accuracy in the harshest of environments. The Pentagon's weather station of choice for tanks.

What's not. Very expensive. No display unit. Requires a computer hookup for data retrieval. Doesn't come with rain collector.

★★ *Runner-up #1*
Capricorn

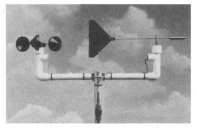

The Capricorn weather stations have been a favorite of professional and industrial users for years. The models have a reputation for reliability, and the display allows you to view several weather functions at a time.

Hinds Instruments, the manufacturer of the Capricorn, offers two models: the Capricorn II ($1220) and the Capricorn II Plus ($1762). Both units monitor wind, temperature, and barometric pressure. The Capricorn II Plus adds humidity, albeit at a substantial extra cost.

Hinds advertises that the Capricorn uses state-of-the-art sensor technology. For example, the humidity sensor uses a "bulk polymer, resistor sensor," which certainly sounds high tech. The wind sensors, however, could use a boost in the accuracy department—above 20 mph, the reading can be as much as 5% off.

A Display from the 1970s

While the sensors may be state of the art, the display console sure isn't. Although you can view temperature, barometric pressure, and wind speed and direction simultaneously, the display panel, with its old-style LED readouts, is not as attractive as units offered by competitors. The console itself is rather sizable, measuring over 15" wide and 7" high. There is a trend indicator that shows whether the barometer is rising or falling, and the display unit can tell you the highs and lows for each function since the last reset. In addition, even though Capricorn bills itself as an electronic weather station, you still have to set the alarm function with manual toggle switches.

Computer Compatibility

Given how expensive the Capricorn weather stations are, I was surprised to learn that a computer interface isn't included. In fact, a RS-232C interface and data formatter run a pricey $245. Plus you need Hinds' WeatherMaster software (another $99) to record, store, and analyze the data.

I saw a printout from the WeatherMaster software and

wasn't overly impressed. While you can use the program for the functions listed above, you'll have to transfer the data to another program for graphing or plotting.

The Bottom Line
If your budget doesn't allow for a Mesotech automatic weather station, the Capricorn might be a good second bet. The rugged sensors have proven reliability in many professional and industrial settings—and Hinds offers a two-year warranty.

Information Box
CAPRICORN II AND CAPRICORN II PLUS
Rating: ★★
List price: Capricorn II is $1220 and the Capricorn II Plus (which adds a humidity sensor) is $1762.
Company: Hinds Instruments, 3175 NW Aloclek Dr., Hillsboro, OR 97124-7135. (503) 690-2000.
Best price in a catalog: The Northeast Discount Weather Catalog (800-325-0360) sells the Capricorn II for $1099.
What's Cool. Display shows several functions at once. Reputation for reliability. Two-year warranty.
What's Not. Pricey, considering computer interface is not included. No rain collector or monitoring capability. The wind sensor mast and mounting kit is additional $49 to $64. Old-style display unit could use face-lift.

★★ *Runner-up #2*
Nimbus Weather Instruments

The Nimbus line of weather instruments enjoys a well-deserved reputation for accuracy. A meteorologist who took a Nimbus barometer and temperature sensor along on an expedition to the top of Mt. Everest raved about their quality. "The two instruments worked beautifully. I would highly recommend them to anyone."

Unfortunately, Nimbus has one drawback: the instruments are not integrated into one weather station. Hence, you have to buy five separate instruments to get the entire weather picture. And that entire picture will cost you: a barometer, a thermometer, and relative humidity, precipitation, and wind sensors will total $2500 or more.

Nimbus' digital barometer ($375) is a typical offering—very accurate and jam-packed with features. Want to know what the pressure was at 7:00 AM? How about Tuesday at noon last week? The Nimbus barometer has the answer, with a 35-day recall of each hour's pressure plus highs and lows. The sensor has a resolution of .03" of mercury—much better than competing products. The display unit can show the rate and direction of change per hour, as well.

The wind monitor ($750) by Nimbus also offers unique features. Besides displaying the wind speed and direction, the unit can show you whether the direction is "backing, veering or steady." Is the wind increasing, decreasing, or remaining constant? Nimbus gives you the scoop, plus displays the peak gust in the last five or 15 minute period.

Museums and other humidity-sensitive environments might be interested in Nimbus' digital remote humidity-measuring instrument. Although expensive ($500), the unit measures humidity with a specially calibrated sensor. As with other Nimbus instruments, the display unit has a 35-day memory and can show the rate of change.

I was less impressed with Nimbus' precipitation monitor. At $300, I was amazed that the monitor doesn't come with a rain collector, which is an additional $272. Or you can use the sensor with any manufacturer's tipping bucket-type rain collector. To its credit, the unit does have two sensors, which provide daily, weekly, or monthly readings. Another neat feature: the monitor gives you the peak rate of precipitation for the past five and sixty minutes.

Recently added to the line of weather instruments is Nimbus' solar radiation unit, which displays either instantaneous or cumulative energy received. The cost: $500 to $600.

Computer Capability? Not Much

Nimbus sells an optional serial port (RS-232C) interface for IBM-PC compatible computers at a hefty price—$50 to $100 per instrument, depending on the number you order. There is no software to help you store or graph the data . . . you're on your own.

The Bottom Line

If accuracy is the most important criterion in selecting weather instruments, Nimbus is definitely worth a look. However, despite the feature-packed display units and high-quality sensors, I was still disappointed that Nimbus has not integrated the instruments into a weather station. The lack of integration impedes the units' usability—most weather stations have built-in alarms if the temperature or humidity falls below a certain value, for example. Nimbus offers no such capability.

Another possible negative: many of the units are battery-operated. This might be a plus if you need weather instruments that are portable, but having to change the batteries every six weeks could drive you to distraction.

Information Box

NIMBUS WEATHER INSTRUMENTS

Rating: ★★

List Price: Each instrument is sold separately, with prices ranging from $300 to $750. A total weather station set-up would run about $2500.

Company: Sensor Instruments Co., Inc., 41 Terrill Park Dr., Concord, NH 03301. (800) 633-1033 or (603) 224-0167.

What's cool. Highly accurate. Fancy features, such as average wind speed and barometric pressure rate of change, add flexibility. Battery-powered units are portable—ideal for emergency management applications.

What's not. High cost. No integrated weather station display—each unit is separate. Lousy computer capability with no software and pricey connections. Battery-powered units means you'll be changing plenty of batteries.

THE THREE R'S: RAINFALL, RELATIVE HUMIDITY, AND RECORD TEMPERATURES

Weather Stations for the Educational Market

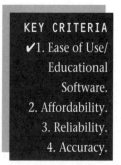

KEY CRITERIA

✔1. Ease of Use/ Educational Software.

2. Affordability.

3. Reliability.

4. Accuracy.

★★★★ *The Best*

Automated Weather Source

The Automated Weather Source combines professional instruments with innovative educational concepts to create a potent classroom learning tool—it's my choice as the best weather station for use in schools.

And this is some weather station. The sensor suite tracks everything you want to know about the weather: temperature (indoor, outdoor, and auxiliary), wind speed and direction, wind chill, dew point, barometric pressure, relative humidity, dew point, precipitation, and even sunshine. You can find out the exact time the high temperature occurred, what the average wind speed is, this month's rainfall, and more.

The digital display is among the best I've seen for any

weather station. You can view nine functions simultaneously, and a large compass indicates the wind direction. I like the trend arrows for temperature, barometer, and humidity.

Computer Compatibility: Among the Best

Hook up the Automated Weather Source to either a IBM-compatible or Macintosh computer, and the fun begins. You don't have to use a computer to monitor the weather with this station, but if you don't, you'll be missing out on some of the best graphics and communication capabilities available today.

The Automated Weather Source is the first weather station to truly take advantage of state-of-the-art computer graphics. The main display screen features creative graphics (thermometer "bulbs" to display temperatures) and informative data (arrows that point to highs and the time they occurred). Easy-to-use software lets you graph and even *map* data, plus it ties into the educational lessons included with the package.

Best of all, your school can "talk" to other Automated Weather Source stations via an ordinary modem (included in the package price). Dial up other schools and view their data. What is the current temperature across town? Your students can instantly get the answer, as well as download historical data from other stations for comparison and analysis.

So what does all this cost? Well, the entire system runs $3350 for schools (or $3950 for a "professional" version) and

includes everything: the sensors, the digital display, a "data logger" (which connects to a computer), communications software, and more. You also get an interdisciplinary lesson packet and software (for elementary, middle, and senior high school curriculums). Basically, everything you need except the computer is in the box, including a modem, cables, power strips/surge protectors, and even an installation video.

If your school can't afford $3350 to spend on a weather station, you may be able to get one *free*. The Automated Weather Source company has formed an innovative public-private sector partnership with local TV stations to roll out weather stations across the country. Local TV stations have teamed with corporate sponsors to purchase Automated Weather Source stations and donate them to area schools. In the Washington, D.C.-Baltimore area, for example, over 120 stations have been installed!

What's in it for the TV station? The free data, of course. By tapping into the Automated Weather Source network, the TV weathercaster can get information from dozens of local stations automatically. How did rainfall vary across town? What were the peak wind gusts in the suburbs? Instead of relying on just one source for data (usually the local airport), meteorologists can answer these questions and improve their forecasting ability, at least theoretically. The station gets "exclusive" data to use in their on-air forecasts and works to improve education at the same time. If you are interested, contact the Automated Weather Source at the number below to see if any TV stations are participating in your area.

The Automated Weather Source network has been so popular (they have 500 stations up and running in 43 markets) that the company has attracted the attention of the National Weather Service and The Weather Channel, both of which are interested in getting access to this data.

The Bottom Line

If you want to teach your students about the weather, the Automated Weather Source is the best solution for the educational market—no other system offers a professional weather station, integrated software with lesson plans, and a communication link to share data with other schools. I can only hope more TV stations will get involved with the program and donate weather stations to schools (or at least contribute to the cost).

What about hobbyists? If you can get past the high cost, the Automated Weather Source would make a perfect home weather station—and the ability to view weather conditions at other stations in your area brings a new depth to weather-watching that might be worth the extra cost.

Information Box
AUTOMATED WEATHER SOURCE
Rating: ★★★★
List price: $3350.
Company: Automated Weather Source,
14115 Seneca Rd., Darnestown, MD 20874.
(301) 258-8390 or (800) 544-4429.
What's Cool. Complete turnkey package includes educational lesson plans and more. Top-quality weather station with fantastic display. Can share data with other schools/stations.
What's Not. High cost. TV stations get the data for free—and some will expect schools to pay the entire cost of the station.

Honorable Mention: Davis Weather Monitor II

Okay, so you don't have the budget to buy an Automated Weather Source. The Davis Weather Monitor (reviewed earlier in this chapter) costs about $600 and enables you to monitor just about everything you want to know about the weather. A software program lets you log data and graph weather information.

Do you have any comments on the weather stations reviewed in this chapter? Call me with your thoughts and opinions at (303) 442-8792.

97

6

More Tools for Predicting the Weather

THERE'S MORE TO OUTSMARTING THE WEATHER than having a fancy home weather station. After figuring out the weather at your home, you also need to know what's happening across town and what's headed your way. This chapter focuses on three low-cost tools that will help you answer those questions: weather radios, scanners, and lightning detectors.

Weather Radios: Bad Weather Turns Them On

As a line of thunderstorms approaches your home, you wonder, "Is this going to be a garden-variety shower or a severe thunderstorm with hail and high winds?" Unfortunately, you can't always tell just by looking—a line of light showers may mask another line of dangerous storms just behind it.

What's the solution? You need a weather radio so you can listen to the latest storm warnings and weather forecasts. The National Weather Service broadcasts continuous forecasts and other vital weather information 24 hours a day from most of its local offices (about 380 locations across the country).

Unfortunately, you can't pick up these broadcasts on a standard radio—you need a special receiver that can hear high-band transmissions. A regular FM radio can pick up 88 to 108 megahertz; the National Weather Service's radio network broadcasts on seven frequencies at around 162 megahertz. Although marine radios, scanners, and the like can receive high-band broadcasts, the least expensive option may be a weather radio. Several of the best weather radios are reviewed later in this section; most cost between $30 and $60.

Most weather radios have unique features that differentiate them from other high-band receivers. For example, many weather radios have an "alert" function that automatically sounds an alarm (a loud ten-second tone) or turns on the radio during severe weather. How does this work? The National Weather Service sends out an alert tone on its radio broadcast when it wants to issue a severe weather watch or warning. After the ten-second tone, the weather service will broadcast the special information. Most weather radios have a built-in feature that "hears" the tone and activates an alarm or (even better) automatically turns on the weather broadcast. Either way, you're alerted to severe weather before it happens.

There is one shortcoming to weather radio broadcasts—the National Weather Service's transmitters are designed to cover only a 40-mile radius. Hence, there are pockets of the country that can't receive these vital broadcasts and warnings. This blind spot proved deadly when a killer tornado destroyed a church in Alabama in 1994, causing dozens of fatalities. The area in rural Alabama was outside the range of the nearest National Weather Service radio transmitter. Shortly after this tragedy, the government announced plans to add additional transmitters by 1995 to cover the 5% or so of the U.S. that currently can't receive severe weather broadcasts.

I should also note that some National Weather Service locations do not broadcast 24 hours a day (for example, the weather radio station in La Crosse, Wisconsin). However, the broadcast hours are extended during severe weather.

How do you know what frequency your local National Weather Service office uses? Besides calling the office directly, you can also get this information from radio manufacturers—for example, Maxon (see review later) includes a chart of all the weather radio frequencies with all its radios.

What You'll Hear

The National Weather Service's broadcasts are hardly what you'd call scintillating radio. Basically, it's a meteorologist who drones on in a monotone, with pacing that makes Paul Harvey sound like a speed reader. I've always wondered what the weather radio would be like if the government hired a silver-tongued FM disc jockey like Casey Casem to do the weather:

> *Yes, kids, it's going to hot! hot! hot, today! Highs in the 90s! Yeow! Get ready for some scattered showers tonight, and those lows will caress the 70 degree mark. Remember, keep your feet on the ground and keep reaching for those cumulus clouds!*

Well, maybe that wouldn't really be an improvement. Anyway, despite the lackluster production values, there is some useful information on the weather radio. Here's a wrap-up of what you'll hear:

1. CURRENT AND REGIONAL CONDITIONS. Every hour, you'll hear the current conditions at your local weather service office (temperature, cloud cover, humidity, wind, and barometric

pressure). You'll also get a summary of regional conditions within a 100 miles or so, with temperatures and wind conditions for towns nearby. I like this part of the broadcast for several reasons—the current conditions give you a snapshot of what's going on in your area. For example, has a cold front passed a town to the north of you, dropping the temperature and shifting the wind direction? Also, you can check the accuracy of your home weather station by seeing what the National Weather Service's local office is reporting.

2. FORECASTS. Weather radio provides continual forecasts, including short-term (36-hour) and long-term (five-day) outlooks. Depending on your location, you may also hear regional or "recreational" forecasts for a specific area or for the entire state. This may include specific forecasts for boaters, skiers, and other sports enthusiasts. In Colorado, we hear the forecast for the entire state—this is quite useful, especially in the winter. Listeners can hear if a Pacific storm is sweeping into the state, for example, and the snowfall estimates give an early indication as to how powerful the storm will be.

3. RADAR UPDATES. If there is any precipitation in your area, the weather radio will give you a radar update, indicating what is happening (light rain, heavy thunderstorm, moderate snow, etc.), the general area of coverage, and the direction of movement. I have to admit I find this information only marginally useful. As they say, a picture is worth a thousand words. Radar "descriptions" over the radio can't compare to seeing a television picture of the radar. In the next few chapters, I'll talk about affordable ways to get the National Weather Service's radar pictures, either on your television or home computer.

4. SEVERE WEATHER AND SPECIAL WEATHER STATEMENTS. When the weather gets nasty, the fun begins on weather radio. As mentioned above, the National Weather Service broadcasts a special tone that activates (or sounds an alarm on) most weather radios. After the tone, you'll hear a message indicating one of the following alerts.

• *Watches.* A severe thunderstorm watch indicates that conditions exist for the formation of severe thunderstorms, which may contain heavy rain, damaging winds, hail, and dangerous lightning. The watch area (depicted on weather maps as a large yellow box) may cover dozens or even hundreds of square miles. Note that although a watch means conditions *may* exist for the formation of severe storms, no storms may have actually formed yet.

Tornado watches are similar to severe thunderstorm watches, except for the additional twist of possible tornadoes. Despite possessing enough technology to obliterate the planet, we still don't have the ability to pinpoint exactly when and where tornadoes will strike. The best the National Weather Service can do is give a general area in which tornadoes *might* form. In Chapter 10, I'll discuss the ingredients for tornadoes in more detail.

☉ *Insider's tip:* The folks at the weather bureau often make severe thunderstorm or tornado watches a complicated lesson in geography. "The National Weather Service has issued a severe thunderstorm watch along and 60 miles either side of a line 30 miles northeast of Little Rock, Arkansas, to 55 miles southeast of Memphis, Tennessee. The following counties are included in the watch area . . ." Excuse me? Unless you have to a CD-ROM map of the U.S. in your brain, figuring out whether you're under the gun for severe weather can be a mind-bending exercise. My advice: get a good county map of your state (or surrounding area) and keep it near the weather radio. This is helpful since the names of counties often have no basis in reality. Take Texas, for example. While Austin is the capital of the state, Austin *County* is actually closer to Houston than Austin, which is in Travis County. Confused yet?

• **Warnings.** In contrast to watches, which only indicate that severe weather *might* happen, as you might expect, warnings are more serious. A severe thunderstorm warning means a severe thunderstorm has been confirmed (by ground observations or radar) and is threatening a particular area. Similarly, a tornado warning means a tornado has been sighted (or indicated by radar). Warnings tend to be short in duration (an hour or less) and confined to a small-scale area such as a county or possibly a town or suburb.

Quite frankly, I think the National Weather Service goofed when it picked the terms to describe these different conditions. The words "watch" and "warning" are too similar and often lead to confusion. People hear the words "tornado watch" and start to panic, even though no warning has been issued. While a certain amount of caution is helpful, I think better nomenclature would clarify the situation. My suggestion: let's replace "watches" with "advisories" for severe weather. In reality, that's what's happening—the weather folks are advising that severe weather is a possibility. By putting the words "severe thunderstorm" or "tornado" *after* the word "advisory" (i.e., an advisory for severe thunderstorms), there might be less undue panic.

As for "warnings," I'd replace them with "emergencies," such as in "the National Weather Service has issued a TORNADO EMERGENCY" for a certain county. Sounds more urgent, doesn't it?

• **Winter weather warnings.** It's also worth noting that the winter season has its own special brand of severe weather. A "winter storm watch" is usually posted several hours (or even a day) in advance of a storm that could dump significant snow or ice on a region. A "winter storm warning," much like other weather warnings, means a storm is already in progress and may result in significant amounts of snow or ice, making travel hazardous.

Notice the key word in the last sentence was "significant." How much snow is "significant" varies from region to region. Southern parts of the U.S., which see little snow in the winter, designate a storm that has the potential to drop just two to four inches of snow as requiring a "winter storm warning." Meanwhile, in the Rocky Mountains, it might take six to 12 inches of potential snowfall to qualify for a warning. What about two to four-inch storms, which are commonplace in the winter in mountainous areas? The National Weather Service is likely to issue a "snow advisory" in such cases, apparently assuming that snowfall that would panic Atlanta won't faze a city like Boise.

If significant winds (gusts or sustained winds over 30 miles per hour) with heavy snow are expected, the National Weather Service might issue a blizzard warning. A winter storm with freezing rain (liquid rain that freezes on contact with the ground, creating a treacherous glaze of ice) may trigger a freezing rain advisory or warning.

All of these varieties of watches and warnings are aired on the National Weather Service's radio broadcasts, much like spring severe weather warnings.

Well, Nobody's Perfect

As you might imagine, the National Weather Service doesn't have a perfect batting record when it comes to warning the public about severe weather. Occasionally, a watch or warning will be canceled a short time after it's issued if severe weather fails to develop. More dangerous, of course, is when the National Weather Service fails to issue a warning for a severe storm. While it's hoped that improvements in weather radar and satellite technology will help spot more storms before they turn deadly, you'll have to expect that the weather service will miss as many as 10% of storms in the foreseeable future.

I've discovered the weather service is more likely to blow a forecast with severe winter weather. This is probably because the dynamics of predicting a major snowstorm are more complex than for a thunderstorm—radar may not detect snow as easily as a heavy downpour or large hail. Low-pressure systems that cause blizzards can move erratically, dissipating in one area only to reform in another. Conversely, a line of severe thunderstorms tends to move in a certain direction at a fairly predictable speed. Nevertheless, while I take the National Weather Service's winter weather watches/warnings with a grain of salt, I'm still grateful for the several hours' (or even day's) warning of such a storm.

Despite the flaws, it's worth investing in a weather radio. You'll usually get hours of warning in advance of bad weather. Even if there is just 15 to 20 minutes of warning for a tornado, this is ample time to take cover.

The Best Weather Radios

You have two choices when it comes to most weather radios: alert or no alert. Basic models just receive the National Weather Service broadcasts; there is no alert function that sounds an alarm or turns on the radio in case of severe weather. Because of this shortcoming, I wouldn't recommend these units as your only weather radio. To help point out which models do or don't have the alert feature, this section is divided into "alert" and "basic" models.

Where do you buy a weather radio? The most widely available models are from Radio Shack. The other models are available from mail-order catalogs (their numbers are noted below).

Most weather radios can pick up the three most common of the National Weather Service's seven broadcast frequencies. These three frequencies are used in 98% of the country. The other four frequencies are used in only 18 locations: Lindsay, California; Ft. Collins, Colorado; Marion, Illinois; Caribou, Maine; Hermitage, Missouri; Marlboro, Vermont; Park Falls, Wisconsin; and three towns in Georgia (Baxley, Valdosta, and Waynesboro), two in Indiana (Bloomington and Marion), and six in West Virginia (Beckly, Gilbert, Hinton, Moorefield, Spencer, and Sutton). If you live in or near any of those locations, you need to call your local National Weather Service office and ask for the local weather radio frequency. Then check to make sure the weather radio you buy can pick up that frequency.

Most weather radios have a small telescoping antenna. This is more than adequate for receiving broadcasts if you live in an urban area or within 50 miles of a transmitter. The reception is usually excellent in most homes or businesses. If you live in a rural area, however, you may need an external antenna to receive the broadcasts (the models that have an external antenna jack are noted below).

How many weather radios do you need? I recommend at least one weather radio with an alert feature for both your home and work place. Locate the weather radio at home in a place where people gather (the kitchen or family room). If you live in an area with a greater risk of tornadoes (such as the Great Plains or Midwest), you may want an additional weather radio in your bedroom since severe weather can strike any time of the day or night. Several companies make portable, battery-powered models—perfect for boaters, hikers, or commuters.

The Ratings
★★★★ EXCELLENT—*my top pick!*
   ★★★ GOOD—*above average quality.*
   ★★ FAIR—*could stand some improvement.*
   ★ POOR—*yuck! could stand some major improvement.*

Alert Models

Maxon ★★★★ The best weather radio on the market is the WX-70 by Maxon. It also happens to be one of the more expensive, retailing for about $60. The Maxon has a feature that automatically turns on the weather service's voice broadcast in case of severe weather. Another plus: the unit has an antenna jack—in case you live more than 50 miles from a National Weather Service transmitter, you can easily add an antenna to pick up broadcasts. The WX-70 model also can receive all seven of the weather radio frequencies and runs on AC power or a nine-volt battery. Available from the Wind and Weather catalog, (800) 922-9463. You can contact Maxon by calling (800) 922-9083 or (816) 891-1093 or write to PO Box 20570, Kansas City, MO 64195.

Midland ★★★ Midland makes two models: the 74-102 ($30) and the 74-105 ($50). While both receive weather broadcasts, the 74-105 can also pick up regular AM and FM radio as well. I suppose this feature is helpful if you want to turn to an all-news or talk station to get additional information about a severe storm. Like the Maxon, the Midland models have a feature that turns on the voice broadcast (instead of an alarm) when severe weather threatens. In a review of this model in *Consumer Reports* (September 1993), the magazine noted the 74-102 "has a timer that's supposed to turn off the alarm after five minutes. Annoyingly, it also turned off the voice broadcast after a few minutes if no alert had been sent." Oh well, at least it has an antenna jack and operates on AC power or a nine-volt battery. I haven't seen the 74-105 in any catalog; the Sporty's Preferred Living catalog (800-543-8633) carries the 74-102 for a whopping $50.

Radio Shack ★★ Radio Shack makes not one but six weather radio receivers. Only two have an alert function (the four non-alert models are reviewed separately below). These two have a unique feature: a "lock" option sounds the National Weather Service-activated alarm until you silence it (helpful if you're out in the garden and didn't hear the original alarm tone). The 12-240 ($40) is a table-top model that works off of household current or a battery, and the 12-143 ($30) is a portable, battery-operated unit. Radio Shack's weather radios have a couple of disadvantages. First of all, there is no option that automatically turns on the voice broadcast—all the radio does is sound the ten-second alarm; you have to manually turn on the voice broadcast. Another drawback is the lack of an antenna jack with these alert models, a major negative if you live in a rural area with poor radio reception. I've used the portable model for a year or so and found it to be functional. My biggest gripe is that the thing eats batteries—three AA batteries only last for a month in the "alert" mode. Since there is no AC-power jack, you're stuck with having to buy 36 batteries each year. While the 12-143 would be a decent choice for a car or "on the go" activity, I wouldn't recommend it as a primary weather radio for your home or work place. Consult your phone book for a Radio Shack near you.

WeatherOne ★★ This rugged portable weather radio ($30) has a retractable antenna that pivots—perfect for hikers, boaters, or other folks who are on the go and need weather information. An alert function sounds an alarm during severe weather, and the unit only works on batteries (one nine-volt, not included). Available from the Safety Zone catalog (800-999-3030) and the Wind & Weather catalog (800-922-9463).

Basic Models

Sony ★★1/2 Sony makes an interesting clock radio (model ICF-C503) that is designed to be mounted under a cabinet, making it perfect for the kitchen or office. In addition to standard AM and FM broadcasts, the radio can also receive weather and even television broadcasts. A clock also doubles as a countdown timer, and the radio even has 20 station-memory presets. It is available from the Hammacher Schlemmer catalog (800-543-3366) for $59.95 (item #53121H). Interestingly enough, the same catalog sells another clock radio that also receives weather radio broadcasts (it's not a Sony, however). Item #53128H combines an AM/FM, alarm, and a large LED clock display with 2-inch numbers. The clock has a snooze button, sleep feature, and an alarm whose volume slowly increases with time. You can even set the clock to wake you up with the National Weather Service broadcast—a scary prospect indeed.

Radio Shack ★ As mentioned earlier, Radio Shack makes four weather radios without an alert function. The famous weather cube (model 12-239, $30) features one-touch weather information and runs off a nine-volt battery. The least expensive

106

Radio Shack weather radio is the 12-242, a battery-operated compact unit that costs $20. If you want a desktop model, Radio Shack offers a couple options. The 12-241 can pick up the three basic weather radio frequencies (like all Radio Shack weather receivers) and costs $22. In addition, the more expensive desktop model (number 12-243, $40) can pick up all seven weather broadcast frequencies and even has an external antenna jack. While these models are nice, the lack of a severe weather alert function limits their usability. Why Radio Shack doesn't make a seven-channel weather radio with an external antenna jack *and* the alert function is beyond me. Consult your phone book for a Radio Shack near you.

Weather Warnings by TV?

After reading about weather *radios*, astute readers may be asking, "This seems silly. Is this the 1800s? Aren't there any similar warning systems for the television?"

Well, a few cities actually do have weather warnings delivered by television. Or cable, to be exact. The cable television company for Plano, Texas, has a weather warning system that is quite amazing. When a tornado warning is issued for Collin County, a special signal from the cable company triggers the converter box on top of the television. This converter box sounds an alarm, turns the TV on, and displays a red screen with the latest severe weather information. If the set is already on, the red warning screen is displayed on all channels. Now, that's impressive.

Every cable company in "tornado alley" (the area from Texas north to Kansas that sees the greatest number of tornadoes) should offer this service.

Scanners: Tuning in Your Town

While a weather radio sounds like a natural tool for weather enthusiasts, a police scanner might seem like a stretch. What good can a scanner do you, besides hearing law enforcement officers talking about taking a donut break? As it turns out, it is a very effective tool for determining what's happening with the weather. Think of all those police, fire, and other municipal employees as dozens of weather spotters. When the weather turns nasty, the radio scanner heats up with these folks relaying information to one another.

Scanners run $120 to $400—not a small investment. But once you listen in, you'll be hooked. Before we review several models of scanners, let's take a look at how you can use a scanner to outsmart the weather.

What You'll Hear

A scanner picks up much more than just police chatter. Here's just a piece of what could be happening:

• *The State Patrol.* During severe winter weather, I listen to a scanner to pick up state patrol officers talking about road closures. There's usually fascinating chatter about which roads are treacherous and which are not. Occasionally, you'll hear a call to sand trucks or snowplows. A scanner is an invaluable tool to find out up-to-the-minute information on road conditions in your local area.

• *Fire departments.* Fire response personnel get called out for much more than fires. During windstorms, overhead electrical lines might rub against trees (or homes), causing sparks and possible fires. In other cases, fire departments are asked to assist with weather-related car accidents or damage incidents. With a scanner, you'll hear instant damage reports when severe weather strikes.

• *Weather spotters.* The National Weather Service operates a network of trained weather spotters who ferret out severe weather. While some of their reports can involve rather technical meteorological terms, the spotters provide information about severe storms. Call your local National Weather Service to find the spotter radio frequency near you. As a side note, most scanners can also pick up the National Weather Service radio broadcasts, just like the weather radios reviewed above.

• *Marine reports.* Scanners can eavesdrop on ship-to-ship and ship-to-shore transmissions, as well as transmissions by the Coast Guard, Navy, and other official departments.

• *Cellular phones.* Here's one of my favorite uses of a scanner. Now, I should point out that intercepting cellular phone calls is illegal. However, there are a couple of loopholes. Some scanners still receive the 800-megahertz frequency used by many cellular phones. And when some cellular sites get overloaded with calls, they use ordinary ham radio repeaters to complete calls and thus are fair game for interception (see below for more information on frequencies).

What's so fascinating (weather-wise) about cellular phone calls? All those phone users unknowingly become weather spotters when the weather turns severe—most report on the weather and road conditions to the people they're calling.

By the way, the controversy over the legality of using a scanner to listen to cellular phone calls may soon become moot. Many cellular companies are switching from analog to digital technology—since a scanner can't pick up digital transmissions of voice conversations, this valuable weather link will soon disappear.

• *The Media.* Ever notice how many television stations have helicopters and mobile satellite units? They dispatch these toys to cover even the most minor storm or weather event. It's fun to listen in on the radio transmissions between the station and the helicopter/satellite truck to see where they're going. You can learn about damage and other weather events *before* you see it on television.

• *And, yes, the Police.* Listen to a police radio frequency, and you'll be amazed how many calls are weather-related. A tree knocked over by a wind gust is blocking a road. Sleet and freezing rain have turned a major thoroughfare into a skating rink. A creek swollen with flood waters is threatening an apartment complex. Even when the weather isn't threatening, it still seems to pop up frequently as a topic of conversation on the police radio.

What to Buy

Radio Shack is probably the most well-known (and easiest to find) source for scanners. On a recent visit, I saw several models that ranged in price from $120 to $400. Base models run $139 to $400, while mobile (or hand-held) scanners cost $119 to $350.

The more expensive scanners have more programmable channels (you have to program each frequency you want to monitor into a "channel"). While this is somewhat of a hassle, in the end you get a scanner that's tuned to your local radio users. How many programmable channels do you need? First, consider that the police (and most other departments) use more than just one channel. Our local police department has five channels. In addition to a channel that dispatches officers, other channels are for officer-to-officer communication. The same goes for fire, ambulance, state patrol, and other emergency response departments. Even our local university's police department has three channels.

As you can see, the channel "real estate" on your scanner can go fast. I recommend a scanner that has at least 100 channels. Weather buffs in densely populated areas might go for the 200 channel models. For example, Radio Shack sells a hand-held mobile scanner with 200 programmable channels for $350.

Another nice feature: "hyperscan." This option lets a scanner zip through 100 channels in one second. That way you won't miss a call, even though you've programmed in a number of channels.

If you'd like to tap into cellular phone calls (for legal purposes only, of course), make sure the scanner can receive the 800-megahertz frequency (the most common band for cellular phones). A hand-held scanner that receives the 800-megahertz frequency with 50 programmable channels costs about $200 at Radio Shack. As mentioned above, you may be able to pick up cellular calls on ham radio repeaters (about 50 megahertz to 150 megahertz, with some repeaters in the 450 megahertz range). As a side note, Congress recently banned the manufacture of scanners that can receive cellular phone frequencies in the 800 megahertz range—as a result, these scanners may become increasingly difficult to find.

Should you buy a base model or a hand-held scanner? I vote for the hand-held scanner. The added flexibility of mobile use wins out every time. During the week, I operate the hand-held off of AC power in our office. Then, if bad weather threatens to wash out a weekend outing, I take the battery-operated scanner along in the car.

So, how much does a mobile hand-held scanner with all the bells and whistles cost? Well, Radio Shack sells a model for $350 that has 200 programmable channels, receives the 800-megahertz frequency, and has "hyperscan." I should note that Radio Shack frequently changes model numbers for scanners—call a store near you for information on the latest offerings.

Frequency Finders

Now that you're convinced a scanner is a great tool for weather buffs, you're faced with the task of finding all the frequencies in your local area. Luckily, Radio Shack has thought of everything. The company sells a book titled *Police Call* for $10 that gives you the answers. The book is published in nine volumes, each of which covers a different group of states. You get all the frequencies for the state, county, and local emergency response departments.

Want more? Want to know the frequency that your local McDonald's uses to take drive-through orders? Or the security detail for a sports arena? Radio Shack also sells *Beyond Police Call* ($10) for all those funky frequencies you never realized you could listen to. Radio Shack even carries more publications with marine and aeronautical frequencies for radio buffs that can't stop scanning.

Lightning Detectors: Seeing the Storm Before It Sees You

What's the most deadly weather phenomenon in the United States? If you think it's tornadoes or hurricanes, guess again. Surprisingly, lightning is the most dangerous, killing 100 people each and every year in the United States. And lightning probably also kills a similar number of computers, cordless phones, and other electronic equipment in its way.

Given how dangerous nature's fireworks can be, getting advanced notice of lightning storms would not only save lives but also property. Unfortunately, while a home weather station might give clues that a thunderstorm is approaching (falling barometer, rising humidity, gusty winds, etc.), none have built-in lightning sensors. And all the National Weather Service's fancy tools, such as Doppler radar, aren't much help either—lightning often appears before the first drop of rain hits the ground.

Fortunately, there are a couple companies that make affordable lightning detectors, ranging in price from $30 to $375. The best are reviewed below.

Who Needs a Lightning Detector?

Quite a few businesses need early warning for lightning. Golf courses, yacht clubs, construction sites, and farms all need a lightning detector to protect against injuries and damage. Moreover, if you have a home office with an expensive computer or other electronic equipment, you might want to invest in an inexpensive lightning detection system. When a lightning storm hit a utility pole near our house last year, the resulting electrical surge zapped our copier (even though it was connected to a surge protector). The repair bill was over $200. If we had had any advance warning of the storm, we might have been able to unplug the copier (and other sensitive electrical devices) before the lightning wreaked havoc with the circuit boards.

Which part of the country sees the most thunderstorms? Florida and the Front Range of the Rockies (Wyoming to New Mexico) probably record the most thunderstorms during the year, with the Great Plains (from North Dakota to Texas) close behind.

How Do They Work?

Ever listened to an AM radio during a thunderstorm? Every time a flash of lightning illuminated the sky, you probably heard a loud crackle on the radio. The same electrical impulses that interfered with your radio's reception can be picked up by the specialized electronics inside a lightning detector. The intensity of the impulse will give a clue as to how close or far away the lightning strike is.

The Ratings

★★★★ EXCELLENT—*my top pick!*

★★★ GOOD—*above average quality.*

★★ FAIR—*could stand some improvement*

★ POOR—*yuck! could stand some major improvement.*

Storm Alert ★★1/2 This small, battery-powered portable device uses a digital signal to analyze lightning strikes. The LED monitor lights give you a visual reading of the lightning's intensity, while a loud buzzer provides an audio warning. The manufacturer claims the Storm Alert can provide up to fifteen minutes advanced warning before a storm.

Available from Edmund Scientific's catalog (item # R39,280 for $169) at (609) 547-8880.

Stormwise Lightning Alert ★★★ Manufactured by McCallie Corporation (PO Box 77, Brownsboro, Alabama 35741-0077; 205-776-2633), the Alert system comes in several models ranging in price from $29.95 to $375. The basic model (LSU-222A) is available directly from McCallie, while the other more expensive models are sold through weather catalogs, as noted below.

The low-cost LSU-222A ($29.95) is a fully weatherproof lightning sensor that alerts you to approaching cloud-to-ground lightning strikes long before you can see the lightning or hear the thunder. A three-way alarm system can be set to beep with each strike or with each "lightning burst" (a huge outbreak of lightning with 40 or more strikes per second).

The LSU-222A is sort of a "do-it-yourself" lightning detection system. Although the unit comes with a buzzer alarm, resistor, and instrument sheet, you have to supply the cable, an outdoor mounting pole, U-clamp, ground wire, ground rod, on/off switch, and nine-volt battery. Fortunately, the detector uses no battery power unless lightning activity is sensed. (As an interesting side note, the company just debuted a new model—the 222B—that is designed for teachers and includes 20 educational classroom lessons.)

If you're not a do-it-yourselfer, McCallie also makes several models with all the trimmings. For example, the LD-8000 (available from American Weather Enterprises catalog, 215-565-1232) costs $149.95 and includes the sensor and a detection meter that displays distant and near strikes on a 1 to 15 scale. Like the less expensive system, this model also has a LED light warning system and buzzer alarm. This system comes with cable, ground wire, and rod.

Computer nerds will like McCallie's software program capable of recording, counting, and graphing over 100 lightning detections per second. The cost: $39.95. The program works with any IBM-compatible PC running MS-DOS 3.1 or higher. Demo disks are available for $3.50. Call (205) 776-2633 to order or get more information.

So from how far away can you pick up lightning? At least 100 miles during the day and a whopping 200 miles at night. The sensor is specially designed to not receive any interference from home electronic equipment and other sources.

I was impressed with the McCallie's models. The instructions were somewhat ponderous (including obscure information such as the "antenna input impedance"), but the system doesn't seem that difficult to use. And the company is available to answer questions about installation or operation.

Have you heard of other affordable lightning detection systems? Call me at (303) 442-8792 to share your discovery.

Part Three

cruising the
weather information
superhighway

7

Easy & Fun Day Trips on the Weather Information Superhighway

FIVE HUNDRED CHANNELS. Movies on demand. Your television makes toast. Yes, the hype surrounding the so-called information superhighway is pretty thick. Yet, surprisingly, much of the *weather* information superhighway is already built. Though it may be a while before you'll surf 500 television channels, today you can view satellite pictures, get the current conditions for Timbuktu, or download Doppler radar pictures.

Many of these information sources don't require a computer—a telephone or television set (with cable, of course) will suffice. But if you really want to cruise the weather information superhighway in style, a personal computer would be helpful. And you don't need a souped-up, expensive supercomputer with a gigabyte hard drive and satellite dish—nope, many affordable personal computers (under $1000) can handle weather day-trips. As we travel the weather information superhighway, I'll spell out exactly what equipment you need to access specific weather information.

This section of the book is organized by level of difficulty: in this chapter, we'll look at the easiest ways to get weather information, like The Weather Channel and the "talking yellow pages." Then it's off to more advanced destinations in Chapter 8, such as on-line services with weather services you can dial up on a home computer. Finally, in Chapter 9, the "expert's section," you'll find information on pulling down pictures directly from weather satellites and other fun activities.

And what about the future? While there's already a wealth of weather information available, there are still parts of the highway that remain under construction. We'll take a look at a new interchange (pun intended) the government is building to make weather information easier and cheaper to get.

As your tour guide to the weather information superhighway, I should note that each source for weather information is listed as an exit (isn't that clever?). Fasten your seat belts and remember to keep your hands inside the car at all times. And no feeding the meteorologists.

117

The Weather Channel
What's required: cable television.

What was life like *before* The Weather Channel? I'll tell you—for weather buffs, there were slim pickings. To get the weather, you had to wait until the evening news, when a blow-dried reporter would read verbatim the National Weather Service forecast and do corny jokes about Girl Scout cookies. If you were lucky, the local TV station would break into programming to warn of an approaching storm. Basically, the attitude was "when we're in the mood, we'll give you the weather."

Then came May 2, 1982—the day The Weather Channel was born. Life would never be the same.

I have to admit to being addicted to this channel, which is received in a whopping 55 million households nationwide. At any time, 100,000 people like me are glued to the set to get an isobar and cold front fix.

Now you can pop on The Weather Channel to get a quick temperature reading and local forecast, which airs every five minutes. While you might assume that many Weather Channel viewers just zoom in and out for this information, research by the A.C. Nielsen Company shows that one in five viewers actually watches for 27 minutes or more. I must be one of those 20% of viewers who finds this stuff fascinating enough to wade through all those commercials for Bac-Os (*now, with the taste of* real *bacon!*) and Sally Struthers pitching a home-study course (*see how easy it is to earn your degree in brain surgery in the comfort of your own home!*)

Just for the fun of it, I spent a day at The Weather Channel to get the inside scoop on how they put on this 24-hour-a-day show. While my trip to the Mecca of Meteorology was chronicled in Chapter 3, this section will highlight what I think are the best (and worst) features of The Weather Channel.

The Best Times to Watch The Weather Channel
Watch The Weather Channel for any period of time and you get the feeling the programming wizards there have decided that the only way to interest viewers in weather 24 hours a day is to dazzle them with dozens of features. One minute you're looking at a radar map of a storm pummeling Cleveland and then, zap! It's off to Europe to check in on a heat wave that's toasting Madrid. There seems to be a conscious effort to keep viewers awake by using different map colors to denote various weather conditions, from ice storms (depicted in deceivingly mellow orange) to severe thunderstorms (in a more appropriate bright red).

To make matters even more complicated, the time that

various segments air changes during different times of the day. Looking for a forecast for Rome, Italy? You'll have to catch "International Weather" at 10 minutes after the hour in the morning. In the afternoon, this report, which gives current conditions and forecasts for Europe, moves to 40 minutes after the hour. Then it disappears in the evening hours, only to reappear at night at 53 minutes after the hour. And then on the weekends, there's another schedule altogether. Confused?

There is some rhyme and reason to The Weather Channel, if you get past the Byzantine program schedule that makes airline schedules look sane. Here's my take on the best times to watch:

1. THE TOP AND BOTTOM OF EACH HOUR. That's TV lingo for the beginning and middle point of each hour (for example, 7:00 and 7:30). In a five minute shot, you get a full look at the weather, from current conditions to the extended forecast. The name of this segment changes depending on the time of day (This Morning's Weather, This Afternoon's Weather, etc.).

This segment usually starts with a satellite picture in motion, with the meteorologist pointing out highlights of trouble spots. Next, it's on to current conditions (usually a look at current temperatures for most major cities) and then the surface conditions map. In some cases, I've seen The Weather Channel experiment with adding map features (that is, cold fronts, high-pressure areas, etc.) onto satellite photos. For example, layering satellite photos with radar images of precipitation areas results in a very complete weather picture. Depending on what's happening, the current conditions report may include radar images, starting with a national view and then zooming to regional or local images.

Then, it's one of my favorite features of this segment: the maps in motion. After showing the current position of fronts and weather systems, the whole shebang goes into motion. You see where the fronts are heading and which ones are projected to stall out over the next 12 hours. A projected temperature forecast of tonight's lows and tomorrow's high temperatures is next, as is a forecast of rain/snowfall amounts.

But that's not all. Many times during the day, The Weather Channel sums up the forecast with a five-day extended outlook of projected highs and precipitation areas (scattered storms, heavy rain, light snow, etc.). While you don't see projected frontal positions, you can get an idea if a major weather change is on the way.

Considering all the weather information you get, it's amazing that this segment is only five minutes long.

Immediately after the current conditions report, The

Weather Channel provides a local forecast—hence, you can tune in The Weather Channel at the top and bottom of the hour and get a complete weather fix.

One scheduling note: while this overall weather wrap-up always airs at the top of the hour, the bottom of the hour version only airs during the late morning to early evening time slot (11 AM to 8 PM Eastern Time) as well as during the nighttime slot (11 PM to 5 AM Eastern).

2. THE FIVE-DAY BUSINESS PLANNER. This handy five-day extended outlook gives a quick overview of future weather, although I'm not sure why they call it a "business" planner.

Starting off with the maps in motion, The Weather Channel gives a quick look at the next 12 hours, with projected precipitation areas and temperature forecasts. Then it's off to the extended forecast, with predicted high temperatures for each of the next five days. You can see how a cold outbreak from Canada sends the temperatures plummeting in the Northeast or how a heat wave builds from the Southwest. In the spring and summer, marked contrasts in temperature will signal a strong cold front and possible severe weather as the cold air clashes with warm moist air from the Gulf of Mexico.

I like the temperature forecast because it shows the projected path of a cold front—for example, will we get hit head on and be in the deep freeze for days, or will the front only deliver a glancing blow, with a quick warm-up to follow shortly?

After the temperatures, it's time for the precipitation forecast. Admittedly, this is not as accurate as it could be, probably because of the uncertainty inherent in long-term forecasting. Rain is shown in various shades of green, with the lightest hues for scattered or isolated showers and the darkest colors for the heaviest storms. Possible severe storms are highlighted in red.

Depicting predicted scattered or isolated showers presents a challenge. Such precipitation may be possible in a wide area (say, the entire state of Kansas), yet the probability that any one location in this area will see it remains quite low. The Weather Channel tends to color these whole areas with solid green, giving the false impression that the rain will be widespread. Perhaps the solution to this problem would be to use a textured or patterned shading of green (such as small stripes) to convey the fact that not all of the area will see rain.

3. YOUR LOCAL FORECAST. Airing every five minutes, the Weather Channel's local forecast is probably one of the most valuable services provided.

Interestingly enough, the way the local forecast appears on your television depends on the type of equipment your cable company has. The Weather Channel sends local fore-cast information by satellite to all of its affiliates—cable com-panies use a computer to digest this information and pro-duce the local forecast you see on your TV. There are several versions of this computer—the first model (the WeatherStar 1000) produces very basic, text-only local forecasts on a pur-ple or blue background. Later versions (including the current WeatherStar 4000) produce colorful graphics and maps to depict weather conditions. Best of all, the later version also includes a snapshot of your local radar and shows the move-ment of any precipitation during the last four hours. What your local forecast looks like (and how informative it is) depends on whether your cable company has invested in the latest technology. Unfortunately, my cable company is one of those cheapskates that only use the bare-bones version—we miss out on the local radar and color graphics. Rats.

Anyway, no matter which version of the local forecast you get, The Weather Channel provides another valuable service for its viewers: severe weather warnings. If the National Weather Service issues a severe thunderstorm (or other type of) warning for your area, The Weather Channel interrupts its programming with a red screen that displays the information. In a normal day, there may be 200 or 300 of these warnings that are broadcast on local affiliates—when there's a major outbreak of severe weather, this num-ber can jump to 1500 warnings broadcast in just one day.

A sample of The Weather Channel schedule for September 1994 follows on pages 122-123.

Weather by Phone
Talking yellow pages and other weather phone information lines
What's required: telephone with touch-tone service.

You're planning to fly to Chicago tomorrow. Should you pack a raincoat or snow shoes? Will a storm delay you, or will the sun be out? The answer may be as close as your phone. Many cities have "talking yellow pages," a service provided by local phone companies that gives up-to-the-minute weather information and forecasts. And, best of all, it's free. Our local phone com-pany, US West, offers this service, and it's easy to use. Just look in the middle of your phone book—you'll find a listing of various news, sports, and weather codes. By dialing one number (in our area, it's 303-754-3000) and entering a four-digit code, you get a recording of the information selected.

The weather information available through this service is quite remarkable—the local forecast includes yesterday's high and low and today's forecast, plus an extended five-day outlook. And what about that weather in Miami? It's one of 32 cities for which you can access tomorrow's forecast and

TIME (IN MINUTES PAST THE HOUR)		PROGRAM TITLE

MORNING (5-11AM ET)

:00 & :23	:23 ON M-F only	*This Morning's Weather*
:03 & :28	M-F only	*Good Morning Forecast*
:08 weekends	:10 M-F	*Boat and Beach Report*
:09 M-F	:18 weekends	*International Weather*
:17	M-F only	*Traveler's Update*
:20	all days	*5-Day Business Planner*
:35	M-F only	*School Day Forecast*
:49	all days	*Tropical Update*
:55	all days	*Michelin Driver's Report*

LOCAL FORECAST EVERY 5 MIN. GIVES LOCAL OBSERVATION, ZONE FORECAST, LOCAL RADAR & REGIONAL FORECAST

MID-DAY (11AM-5PM ET)

:00 & :30	except 4PM M-F	*This Afternoon's Weather*
:00	at 4PM M-F	*The Weather Classroom*
:08	exception—:15 after 4PM	*Boat and Beach Report*
:10, :15, :17, :39 & :43		*Weather features*
:39		*Travel Forecast*
:40	moves to 4:53PM M-F	*International Weather*
:49		*Tropical Update*
:55		*Michelin Driver's Report*

LOCAL FORECAST EVERY 5 MIN. GIVES LOCAL OBSERVATION, ZONE FORECAST, LOCAL RADAR & REGIONAL FORECAST

LATE DAY (5PM-8PM ET)

:00 & :30	*This Evening's Weather*
:09	*Boat and Beach Report*
:16	*Heat Wave Alert*
:20	*5-Day Business Planner*
:41	*Traveler's Forecast*
:46	*Tropical update*
:55	*Michelin Driver's Report*

LOCAL FORECAST EVERY 6 MIN. GIVES LOCAL OBSERVATION, ZONE FORECAST, LOCAL RADAR & REGIONAL FORECAST

EVENING (8PM-11PM ET)

:00	*This Evening's Weather*
:14	*Boat and Beach Report*
:16	*Heat Wave Alert*
:22	*5-Day Business Planner*
:43	*Traveler's Forecast*
:50	*Tropical update*
:55	*Michelin Driver's Report*

LOCAL FORECAST EVERY 6 MIN. GIVES LOCAL OBSERVATION, ZONE FORECAST, LOCAL RADAR & REGIONAL FORECAST

NIGHT (11PM-5AM ET)

:00 & :30		*Today's Weather*
:09		*Boat and Beach Report*
:11		*Travel Forecast*
:21 or :22	(start time varies)	*5-Day Business Planner*
:39 & :48	11PM-2AM ET	*Pacific Regional Forecast*
:47		*Tropical update*
:53		*International Weather*
:55		*Michelin Driver's Report*

LOCAL FORECAST EVERY 6 MIN. GIVES LOCAL OBSERVATION, ZONE FORECAST, LOCAL RADAR & REGIONAL FORECAST

DESCRIPTION

Covers weather for NE, Midwest, and South
Forecast of upcoming weather events for next 24 hrs.
Weather relating to boating, beach, & outdoor activities
Conditions across Europe & in Europe's major cities
Reports current and future weather impact
Tells impact of upcoming weather on travelers
Suggests appropriate outerwear for school children
Reviews activity in Atlantic, Gulf of Mexico, and Caribbean
Current and future weather impacting highway travel

Look at national weather, highlighting active weather
Program for teaching principles of weather with live host
Weather relating to boating, beach, & outdoor activities
Taped vignettes relating the weather to daily life and lifestyles
    touching on health, environment, vacationing, racing and baseball
Major cities outlook for business and weekend travel
Conditions across Europe and in Europe's major cities
Reviews activity in Atlantic, Gulf of Mexico, and Caribbean
Current and future weather impacting highway travel

Full look at national weather, highlighting active weather
Weather relating to boating, beach, & outdoor activities
Occurrence of major episodes of heat and/or humidity
Tells impact of upcoming weather on travelers
Major cities outlook for business and weekend travel
Reviews activity in Atlantic, Gulf of Mexico, and Caribbean
Current and future weather impacting highway travel

Full look at national weather, highlighting active weather
Weather relating to boating, beach, & outdoor activities
Occurrence of major episodes of heat and/or humidity
Tells impact of upcoming weather on travelers
Major cities outlook for business and weekend travel
Reviews activity in Atlantic, Gulf of Mexico, and Caribbean
Current and future weather impacting highway travel

Look at national weather, highlighting active weather
Weather relating to boating, beach, & outdoor activities
Tells weather conditions in major cities
Tells impact of upcoming weather on travelers
Provides Pacific coastal states with forecast
Reviews activity in Atlantic, Gulf of Mexico, and Caribbean
Conditions across Europe and in Europe's major cities
Current and future weather impacting highway travel

(based on September, 1994 information)

expected temperatures. And not just for big places like New York and Boston—you can also get the weather for towns like Indianapolis, Salt Lake City, and Seattle. If that weren't enough, information on international weather conditions is available, with forecasts for 12 cities around the world. High and low temperatures are reported in both Fahrenheit and Celsius.

In addition to the weather, you can obtain specialized information on this free information line. Our local service offers snow reports from ski resorts and area road conditions. In beach front communities, tide times and surf conditions are provided.

When I travel, I use this service frequently. One slight negative: while the service is free, you usually are subjected to a brief advertisement before you get the requested information.

If you need weather information that is more detailed than what is available through the "talking yellow pages," The Weather Channel offers an impressive 1-900 service. "The Weather Channel Connection" can be reached at 1-900-WEATHER (1-900-932-8437). Current reports and forecasts for 600 U.S. cities and 225 international locations are available. In addition, you can access specialized reports for boaters, drivers, skiers, and fall foliage watchers, as well as the latest hurricane information. The cost is 95¢ per minute.

What if your phone is blocked from calling 900 numbers? Well, The Weather Channel has thought of everything and now has an 800 service (1-800-WEATHER) that offers the same weather information. You pay by VISA or Mastercard, and the cost (95¢ per minute) is the same as the 900 service. This might be helpful if you're on a road trip and the hotel you're staying at blocks 900 calls—at least you'd have an alternative for getting the latest weather information.

Even the National Weather Service has gotten into the 900-number act. You can call 1-900-884-6622 to hear the latest hurricane/tropical storm information (including positions) as well as forecast information for various U.S. cities. The cost is 98¢ per minute.

Fax It!
Weather information via the fax machine
What's required: a touch-tone telephone and a fax machine.

A traveling salesman is trapped at a hotel in the middle of nowhere, South Dakota. There are dark clouds on the horizon, and he's got to make it to Omaha, Nebraska, by nightfall. He flips on the TV only to discover this is the only hotel in the U.S. that doesn't have The Weather Channel. How can he get timely weather information so that he can plot his trip?

Well, if the hotel has a fax machine, he's just been saved.

Alden Electronics offers an incredible amount of weather information that can be sent to any fax machine in the world. The weather-by-fax service is called ZFX, and it's pretty amazing.

Just dial the company's 800 number, punch in a few codes, and whamo! The weather information you need is zapped to a fax machine. ZFX offers national radar pictures, regional radar views, surface weather maps, forecast charts, severe weather outlooks, snowfall predictions, and much more—all of it is available 24 hours a day.

If that weren't enough, you can also get faxes of satellite images, information for pilots such as upper-level wind forecasts, five-day extended outlooks, and more. ZFX also has data for Canada, Alaska, Hawaii, Puerto Rico, and the U.S. Virgin Islands.

The cost for this service is quite affordable. There is a $7.99 one-time sign-up fee, and each faxed "product" ranges from $1.50 to $2.00. That's it.

Let's return to the poor traveling salesman who's about to get hit with a blizzard. By using ZFX, he can get a fax of a satellite image of the central U.S. (code #3251). The 12 and 24-hour forecast map (#2093) and the real-time weather radar image from Omaha (#3059) would also be of benefit in our salesman's situation. Thanks to all of this up-to-the-minute weather data, he can plan the quickest and safest route to Omaha. Moreover, with a cellular phone and portable fax machine, it's even possible to receive crucial weather information while zipping down the interstate.

For more information, contact ZFX sales at Alden Electronics, 40 Washington St., Westborough, MA 01581-0500. (800)-876-1232 x303.

More Difficult Trails on the Weather Information Superhighway

NOTHING HAS REVOLUTIONIZED WATCHING THE WEATHER more than the personal computer. Affordable software programs let you view wind gusts, graph this month's rainfall, or (with the help of a modem) tap into the rich vein of weather information available on-line.

Take a home weather station and add a personal computer—poof!
Instant weather database

What's required: A Davis Weather Monitor II and a personal computer (IBM-compatible or Macintosh). Call Davis at (800) 678-3669 or (510) 732-7814 for specific computer requirements.

As you read in Chapter 5, my pick as the best home weather station is the Davis Weather Monitor II—and one reason is Davis' affordable software program, Weatherlink ($165). Weatherlink lets you view current weather conditions on your computer, store data on your hard drive, and graph the weather, such as last month's high temperatures or today's wind gusts by hour.

As this book went to press, Davis debuted a revised version of Weatherlink that fixes some of the program's drawbacks. A new, completely revised database format allows you to access historical data faster. You can also download the data from the weather station automatically—before, you had to be there to answer prompts like "download data?" and "clear archive?" Other new features: you can edit data, choose bar and line graph colors, and print to a color printer. Unfortunately, this new version of Weatherlink is only available for IBM-compatible users—Macintosh fans must wait for a similar revision, hopefully in the near future.

Of course, Davis isn't the only company that makes software that will turn your computer into a mean weather machine. Weather Dimensions' (800-354-1117) software program is a powerful tool that creates a strip-chart history of your home weather, complete with current conditions. This software works only with IBM-compatible computers.

One weather station/computer software package I discovered right before this book went to press is the Observer by Beaverton, Oregon-based Fascinating Electronics, Inc. (800-683-5487 or 503-292-5233). This package includes an anemometer and wind vane, barometer, rain gauge, thermometer, humidity sensor, computer interface, and software for $659. (Another version, the Observer Plus, includes three more thermometer sensors for $729.)

What's most interesting about Fascinating Electronics' Observer is their software program—you can view up to five temperatures, rainfall, humidity, barometric pressure, and wind direction simultaneously. Two other unique features are a scrolling graph of wind speed and pop-up trend graphs (for any function, e.g., rainfall, barometer, temperature) that can be moved anywhere on the screen. While I wasn't able to do in-depth research on this product before this book went to press, I hope to have a full report on this station in the next edition of this book.

No matter which program or station you purchase, the greatest benefit provided by these software packages is to be able to see how one part of the weather relates to another. By graphing barometric pressure and temperature, you can see how a cold front passage shifts the trend in air pressure at the same time the temperature takes a dive. By monitoring weather trends, you'll begin to see quirks in your own neighborhood's climate.

For example, I've noticed a fascinating trait to my area's climate. We live on a ridge 2000 feet above and five miles outside Boulder, Colorado. When a weak cold front comes in the backdoor of Boulder (that is, pushes in from the east, instead of the north), there is often a tug-of-war between the two air masses. The slightly colder air tries in vain to push against the warm west winds that regularly buffet the foothills. The result: a see-saw of the barometer and temperature readings—when the cold front pushes past, the barometer rises and the temperature drops. Then, when it loses the battle as the west winds push it back, the temperature recovers and the barometer drops. This on-going drama can continue for several hours before the cold front usually wins.

Brewing up Home Forecasts

What's required: IBM-compatible computer, at least 640K machine with CGA monitor. A 5.25 disk drive is acceptable, but a 3.5-inch disk drive or 20 megabyte hard drive would be better. Also, a thermometer, barometer, and humidity gauge.

Weather instruments are great for telling you what's presently happening with the weather, but wouldn't it be neat if you could enter your observations into a personal computer and have it spit out a forecast?

Keith Haley of K&H Enterprises has created a low-cost software program that does just that: the Weather Pro 5.5. All you do is enter your own weather observations (temperature, barometric pressure, humidity), and the program creates a specific forecast of temperatures and rainfall for the next three days! Weather Pro 5.5 doesn't allow you to connect weather sensors to your computer—you must manually enter your observations (such as the day's maximum and minimum readings for barometric pressure). When you first install the program, you also have to enter some background information on your location (such as the altitude, latitude and longitude, etc.).

I previewed a demonstration version of Weather Pro and found it to be very interesting. My only criticism: I wish it could be linked to my Davis home weather station to download data automatically. Manually keying in data day after day might get tedious after a while.

In addition to creating forecasts, Weather Pro also has safety tips on severe weather, allows you to graph monthly and yearly weather statistics, and even gives sunrise and sunset times.

Besides Weather Pro, Keith has created three other weather software programs. WeatherStat allows you to analyze and graph weather information in more detail than Weather Pro. Keith has also written a reference library program that is essentially a weather almanac on disk. You get historical weather data for 66 cities and information on meteorological books, equipment, etc. Each program is $29.95, or you can buy all four for $89.95. *To order, call (616) 763-3479 or write to K&H Enterprises, 325 Olivet Rd., Bellevue, MI 49021.*

129

Weather On Line
Using CompuServe, America On-Line, GEnie

What's required: A personal computer (IBM PC-compatible or Macintosh) and modem. The modem should have a speed of at least 2400 baud—9600 baud and 14,000 baud are even better.

No matter how much I like The Weather Channel, I have to admit there is one flaw to their rapid-fire presentation of weather information—those maps move too darned fast! Just when you get a good look at a radar image for your area, it seems like the channel zips to the current conditions in Bora Bora. Wouldn't it be nice to download these same satellite pictures, radar maps, and forecasts to your computer so you could digest them at your leisure? Well, there is a way you can do all that and more. Just strap a modem to your personal computer and launch yourself into the world of on-line services.

What can you do once you've logged on? More than you might think. Besides downloading satellite maps and radar images to view on your computer, you can also check the five-day forecasts for a myriad of cities, both domestic and international. Or join a discussion with other weather junkies about the latest tropical storm.

That's just the beginning. Here's a wrap-up of the best on-line services for weather information:

CompuServe
To sign up: (800) 848-8199.
Costs: $8.95 per month for basic service.
Freebies: The basic introductory package for $39.95 includes CompuServe Information Manager (software that helps you navigate the service), one month free access time, and $25 credit toward "extended services" such as research libraries.

Amazing is how I'd describe CompuServe's weather offerings. The sheer volume of weather maps, pictures, and information makes it well worth the $8.95 per month fee. For example, you can view or download radar, satellite, and surface condition maps. Provided to CompuServe by Accu-Weather, these full-color maps are the same ones you see on your local television weathercasts. The current temperature map, which uses color shadings to depict various temperature ranges, is updated every 15 minutes. If you're looking for aviation weather information such as hourly surface observations and winds aloft, CompuServe has got you covered. You can download this data and then use an affordable software program to create maps and analyze the weather (see review later in this chapter).

CompuServe offers updated forecasts (both short-term and extended) for cities and even entire states—a plus if you're planning a road trip. Severe weather alerts, precipitation outlooks, and marine weather round out the service's offerings.

Of course, there's much more to CompuServe than the weather. I like the dozens of "forums," conversation areas organized by topic. For example, the forum for ham radio enthusiasts includes downloads of scanner frequencies and other helpful information.

CompuServe isn't without its faults, of course. Their navigational software program, the CompuServe Information Manager, is a necessity to get around the somewhat confusing network. Even with this graphical interface, CompuServe still lags behind America On-Line (see below) for user-friendliness.

Members also gripe about the extra fees—you can run up hefty bills if you dip into CompuServe's research libraries, and there are extra fees to download some weather maps. Make sure you get information on the latest surcharges before you rack up a big bill.

America On-Line
For more information: (800) 827-6364.
Cost: $9.95 per month for the first five hours. Additional time runs six cents per minute (or $3.50 per hour).
Freebies: The free membership packet includes 10 hours of free access time and connection software.

What's so cool about America On-Line? For starters, check out the interface. A fancy, full-color graphics program makes navigating the network easy and fun. A simple click of a mouse lets you access information, download files, or view messages—much like the easy-to-use Macintosh system.

I cruised the weather section and found a number of interesting items: the latest in weather news, updates on tropical storms and hurricanes, forecasts for U.S. and international cities, and, of course, weather maps.

Weather Services International (more commonly known as WSI—see review in the next chapter) provides America On-Line with many of their maps. Although you can't view the maps on-line, you can download them to your computer. Once you've downloaded a map, you can zoom in on one particular area. The resolution on the maps is just OK; the images are better for a national view than for trying to figure out what's happening on a local scale. For example, I downloaded a radar image and satellite map the night in 1994 when tropical storm Beryl was bearing down on Florida.

The forecasts part of the weather section is worth mentioning—with a few clicks of the mouse you can get a five-day extended outlook for just about any city in the U.S. The forecasts include temperature highs and lows and a precipitation outlook.

131

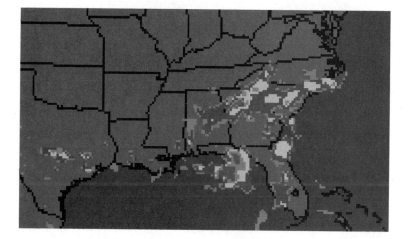

Besides all the forecasts and weather maps, you can engage in discussions on various weather topics. The weather section includes a message board with 50 topics covering everything from which home weather station is the best, to the latest news on tropical storms. For me, this was the most interesting part of America On-Line's weather information—I was fascinated to learn about new weather associations and discount sources for meteorological instruments.

More On-Line Sources for Weather Information

GEnie (800-638-9636) and Prodigy (800-776-3449) are two other on-line services with weather data, albeit in a more limited scope than CompuServe and America On-Line. Prodigy's weather information is skimpy, but GEnie offers current conditions and forecasts for 700 cities in the U.S., Canada, and the Caribbean.

In addition, the U.S. government maintains a free weather database just for pilots, who have an obvious need for accurate and timely weather information. And did you know that anyone can access this information via a toll-free number? It's more than just aviation jargon—the database includes weather information in "plain language," too. You can access this database by using Contel DUAT (DUAT stands for Direct User Access Terminal—don't ask why). Use your modem to dial 800-767-9989. Another company that offers access to this database is Data Transformation Company—their DUAT database includes Accu-Weather products like radar, satellite images, and more for $1 per minute. Call 800-243-3828 for more information (or by modem, 800-245-3828).

I should also note that two universities, the University of Illinois at Champaign and Ohio State University, operate weather bulletin board systems. Each has a series of weather maps you can download, plus discussion groups about various weather topics. To reach these bulletin boards, you'll need to access the Internet, that famous computer network that has received so much attention lately.

So, it's simple, right? You just log onto the Internet? Not so fast, cyberspace breath. The only people who have free access to the Internet are computer users at major research universities. The rest of us (the unwashed general public) must go through other "gateways." America On-Line offers one such gateway. Unfortunately, America On-Line's gateway to the Internet is slow and often crowded. Another option may be Delphi (call 800-695-4005 for information), an on-line service that offers unlimited access to the Internet. Delphi costs $10 per month, which includes four hours of connect time. The only disadvantage to Delphi: the service uses a text-based system that seems primitive compared to the slick graphics of America On-Line. As a result, it's probably not for beginners.

Now what?
Using inexpensive "shareware" programs to massage weather data

What's required: IBM PC-compatible computer with a minimum 286 processor (386 is better). EGA video monitor, math coprocessor recommended, MS-DOS with 512K or more available memory.

Okay, you've signed up for CompuServe and know that you can download current weather observations for the entire United States onto your humble personal computer. But what good does it do you to know that the barometric pressure in Des Moines is 29.77 inches or the temperature in Kansas City is 48°? Luckily, Jim Haywood of Grand Rapids, Michigan, has created an affordable software program called WeatherView ($29.95, including shipping) that turns all this raw data into colorful maps and more.

All you do is download the surface weather data from CompuServe's database of 6000 weather stations in the U.S. and Canada. With a modem that works at a speed of 2400 baud, this takes about six minutes. Then WeatherView goes to work, reformatting the data to make it more usable. You can map rainfall (in one of five colors to indicate intensity), plot temperature, or zoom in on a particular part of the country. The program displays just about any weather parameter, from wind direction to dew point. There are even versions of WeatherView for Canada and Europe. Perhaps WeatherView's only shortcoming is it doesn't have a version (yet) that works with higher resolution VGA monitors.

WeatherView is available from two sources. You can send $25 plus $4.95 shipping to Jim Haywood, 2157 Forest Hill SE, Grand Rapids, MI 49546. Or you can download a demonstration copy of the software from CompuServe (AVSIG forum, library #1, under the name wxview.zip). The demo version has limitations that are removed when your copy is registered with the author.

If you want to go whole hog, check out WeatherGraphix, a low-cost program ($45) that has much the same features as WeatherView. WeatherGraphix was developed by Tim Vasquez, a member of the Texas Severe Storms Association. As a result, his program has become a favorite of storm chasers. *For more information, contact Tim Vasquez at PO Box 9808, Abilene, TX 79607.*

9

Experts Only:
The Most Challenging Journeys
on the Weather Information
Superhighway

Y OU KNOW THE FACES of The Weather Channel meteorologists better than your own family. Your home weather station is so sensitive it can pick up the weather on Mars. You've even ventured on-line to download the latest weather in New Zealand. Now that you've been hooked on the weather, it's time to cruise the "expert's only" section of the weather information superhighway.

What's so challenging about these trips? Do you have to have three degrees in meteorology to understand this stuff? Actually, no college sheepskin is necessary. What you do need in order to access this weather information is *money*. And lots of it.

Many of the following weather sources are expensive, requiring a significant investment in software and specialized equipment. While you can use any personal computer with the weather software and services listed in the last chapter, the ventures described in this chapter will require a fancy IBM PC-compatible computer outfitted with a fast processor chip and a high resolution monitor. I'll spell out the specific requirements in each section below.

Sorry, most Macintoshes won't be able to make the trip. Only a few software programs have Macintosh versions, although the latest Macs (specifically the Power Macintoshes) have the capability to work in DOS and Windows environments and may be able to run these programs.

So, what will you be able to see about the weather in this chapter? Just about anything you ever imagined. First, we'll explore weather databases that are stuffed with every fancy map and piece of data you could want. Next, it's off to the world of weather "facsimile," a little-known way of receiving satellite photos by shortwave radio. Finally, we'll look at the big "Earth stations" that can receive satellite photos directly.

The weather information you can get from these sources comes in two flavors: text-only and graphics. Weather sources with text-only information provide current conditions, fore-casts, severe weather warnings, and so on. While this data is valuable, I have to admit that snazzy graphics make weather-watching more fun. Today you can download satellite pic-tures, radar images, forecast maps, and more on your personal computer. From there, affordable software programs let you zoom in on your local area. You can even "loop" the maps together to animate a storm moving across the radar.

Many of these databases let you download NEXRAD radar images to your computer. NEXRAD stands for "Next Generation Weather Radar," a network of radar stations that eventually will include 160 sites in the United States. This new technology has a much better resolution than what was available previously. While older radars could see just precip-itation (displayed in three colors—green, yellow, and red), NEXRAD can detect wind direction (pinpointing possible tornado development), the rates at which precipitation is falling, and more.

The Weather Databases

8 Ever watched your local television weathercast and wondered where they got all those neat toys? From Doppler radar images showing a storm sweeping toward your town to colorful satellite photos in motion, you might think the television station has a battery of computer experts on call around the clock.

Think again—most stations buy those fancy weather graphics in packages from information providers like you would buy a loaf of bread off the shelf. These weather "gro-cery stores" are three somewhat obscure companies—Weather Services International Corporation (more common-ly known as WSI), Alden Electronics (WSI and Alden are both based in Massachusetts), and Accu-Weather (State College, Pennsylvania), a slightly better-known service thanks to their weather reports on radio stations.

It's not widely known, but all these companies also sell weather information to the public. You read right—you can get the same fancy maps and graphs you see on television downloaded to your home computer. Because these are graphics-intensive files, however, you will need a powerful computer with enough memory and a high resolution moni-tor to make the most of the data.

When you deal with these companies, you'll quickly dis-cover the good news/bad news dilemma of accessing weather data from them. On the one hand, you can get incredible graphics and maps that will boggle the mind. On the other hand, it will cost you an arm and a leg. Many charge hun-dreds of dollars for software and have stiff "connect" charges that run $100 or more *per hour.*

Where do WSI, Alden, and Accu-Weather get their weath-

138

er information? From the National Weather Service, as you might guess. But, wait a minute, isn't that a government agency paid for by your tax dollars? Shouldn't this information be free? Good point—in fact, the government is working on ways to providing the public with all these weather goodies at low (or no) cost. At the end of this chapter, we'll peek into the future of the weather information superhighway.

Meanwhile, these three companies have a virtual monopoly on the delivery of weather information such as radar images and satellite pictures. Of course, most do more than just relay the National Weather Service data verbatim—they "add value" to the data by enhancing it with other information. For example, WSI will take a basic radar image and add severe weather information, wind profiles, and more. This computer "massaging" of data may make it worth the steep cost of accessing the information.

Here's a wrap-up of what WSI, Alden, and Accu-Weather offer the weather junkie:

1. WEATHER SERVICES INTERNATIONAL CORPORATION (WSI)

For more information: 4 Federal St., Billerica, MA 01821. (508) 670-5000.

Quick Summary: WSI offers a software program (Weather For Windows or Macintosh) that lets you automatically download on your IBM PC or Macintosh computer any image or chart from their weather database. A demonstration disk is available.

What's required: IBM PC or Macintosh computer and a modem.
• For IBM PC compatibles, at least a 386X (or greater) with Microsoft Windows 3.1 or higher, 10 megabytes of free disk space. At least four megabytes of memory and DOS version 3.0 or higher. VGA or SuperVGA (256 colors) graphics card and monitor. The SuperVGA monitor is recommended. Modem should be at least 2400 baud (9600 recommended).
• For Macintoshes, you need a computer than can run system 7.01 or higher, at least 4 megabytes of RAM, a high density 1.44 floppy drive, and the "Quick-Time" extension (version 1.5 or higher, to use the looping animation feature).

Wow is about all I can say about WSI's incredible weather database! I just viewed a demo of their program and was floored by the incredible graphics and information available. I played a "radar loop" that showed Hurricane Andrew slamming into Homestead, Florida. Another radar image focused on the southern U.S., layering precipitation echoes (green, yellow, red) on top of lightning data (shown as small purple strikes). And that was just the beginning.

Before I go much farther, I should warn you: this is not cheap. The software alone is $495 (for either Mac or IBM). And then you get hit with the connect charges. Are you sitting down? WSI charges $135 *per hour* to access weather data using a 9600-baud modem. That's a staggering $2.16 per minute—and considering that downloading a simple radar image will take several minutes, you could run up a whopper of a bill very fast.

Cost aside, WSI has been called the "Rolls Royce" of weather databases, and it's an apt description. Basically, any piece of weather information you want is here, including any satellite picture you could need. Want to check out the western coast of Africa for any storms rolling off into the Atlantic, a possible precursor to tropical storms in the hurricane season? It's here, as are satellite pictures for the North Pole, Europe, and even Australia. Just type in your city location, and WSI can send you a satellite picture centered over your location.

The radar images are truly amazing—WSI has all the NEXRAD images, in both "unaltered" and "value-added" versions. The latter are images layered with additional information, such as a storm's movement, the height of cloud tops, an estimation of rainfall rates, and more. You can change the colors of any part of the map, in case you've ever wondered what the oceans would look like in bright fuchsia.

WSI's software includes a helpful scheduler, enabling the user to tell the program exactly what to download and when to do it. For example, every day at 7:30 AM you could download a radar image and satellite picture to check the day's weather. Another interesting product from WSI: a ten-day forecast map! Forget those five-day forecasts you see on TV and become the first to know when a major weather change is on its way.

So, is it worth it? Well, if you have the budget, there probably isn't a better way to get weather information than WSI's Weather For Windows (or Macintosh). You can get a taste of WSI's products by watching The Weather Channel, CNN, or downloading maps from America On-Line.

A spokesperson for WSI told me that the company was considering several new products in the near future that would have lower costs. For example, international weather watchers might be able to access at a reasonable cost only the specific part of the WSI database that interests them. Contact WSI for the latest offerings and prices.

2. ALDEN ELECTRONICS

For more information: 40 Washington St., Westborogh, MA 01581-0500. (508) 366-8851.

Quick Summary: Alden offers an interesting "weather-by-fax" service called ZFX. Other products are quite pricey, including software programs to download radar data and the equipment to receive satellite pictures.

What's required: A fax machine for Alden's ZFX service. For Alden's modem-delivered data and graphics programs, you need a PC/AT-compatible computer, 386 or 486 CPU, two megabytes of RAM, 40 megabyte hard drive, VGA graphics board and monitor, DOS version 3.3 or higher, a parallel port and modem (at least 2400 baud, while higher speeds up to 19,200 are supported).

Alden's most affordable weather information source is ZFX, the weather-by-fax service described in Chapter 7.

If radar is what most interests you, Alden offers "Allrad" software to download a myriad of NEXRAD radar images from their database to your computer. Among the more fascinating products is a map that shows a storm's total rainfall and another image that gives the height of cloud tops (an indication of how severe a storm is). All of the radar images are unaltered and received directly from each radar site.

Like WSI, Alden doesn't exactly give away this information. The software costs $995. You pay a $260 flat monthly fee for unlimited access to the information—there are no per minute charges or other fees.

If money is truly no object, check out Alden's weather satellite receiver system. This $17,000 computer system includes all the computer equipment and antennas to receive just about any satellite picture available.

3. ACCU-WEATHER

For more information: 619 W. College Ave., State College, PA 16801. (814) 234-9601 x400.

Quick Summary: "Accu-Access for Windows" is a software program that lets the user download weather information, images, and maps to a personal computer.

What's required: IBM PC or compatible computer with a 386 processor, VGA monitor, and a graphics card with at least one megabyte of video RAM. Both a DOS and Windows version is available. With this set-up, you can display four graphics at once, with one of those graphics looping through 10 images. With a computer that has more processing power (a 486 or higher), you can have two graphics looping through images simultaneously.

Accu-Weather is probably the best known weather information service in the U.S.; their on-staff meteorologists can be heard giving the weather on hundreds of radio stations each day. So it is surprising that Accu-Weather's offerings for weather enthusiasts lag somewhat behind competitors' products.

I tried an early version of a computer program that accessed Accu-Weather's database and was disappointed. The "Accu-Weather Forecaster" had lousy graphics and skimpy information. Accu-Weather's latest offering (Accu-Access for Windows) at least solves the earlier program's lack of punch: this software program lets you access NEXRAD radar, as well as color satellite images, weather maps with surface conditions, forecasts, and more. Accu-Access also has real-time lightning strike data, hurricane forecasts and reconnaissance reports, and computer forecast models up to ten days, plus marine and aviation information.

The computer interface for Accu-Access isn't as slick as WSI's Weather For Windows—while you can pan, zoom, and loop maps together, the icons and commands to do this aren't as intuitive as WSI's program.

In addition, I found Accu-Weather's customer service to be lacking. I called the company six times (and left six messages) to get information on their database. On the seventh try, I finally got in touch with Terry Lindquist, a company spokesperson who provided cost information and other background on Accu-Weather's database.

The basic cost for Accu-Access runs 10¢ to $1.21 per minute, depending on the baud rate of your modem and the time of day you call. For example, accessing the database from 5 AM to 7 PM with a 9600-baud modem would be billed at the maximum rate ($1.21 per minute). The software costs $14.95, plus shipping.

What's the most popular product on the Accu-Weather database? NEXRAD radar images probably would take the prize. Terry told me that Accu-Weather creates special regional and national radar "mosaics" and adds other information to the images. It takes about 20 to 30 seconds to download a NEXRAD image at 9600 baud. A close second to NEXRAD in popularity: hurricane information. Accu-Weather provides all the latest hurricane information and includes special satellite loops and radar images on the database when a tropical storm forms in the Atlantic.

I should note that Accu-Access for Windows is scheduled for an overhaul that will be released in 1995. The updated version will have more of a "point and click" interface with buttons for radar images, satellite pictures, and more (similar to WSI's software; see review above). Unfortunately, the current version of Accu-Access requires quite a few keystrokes to download data.

I must give Accu-Weather a mixed review. While the connect charges are definitely more affordable than WSI's Weather for Windows, the software's weak interface tempers my recommendation. And the customer service was a major disappointment, especially for a company that bills itself as the "world's weather leader."

4. Weather Services that Won't Break the Bank: WeatherBank & The Weather Network.

One of the most affordable sources for weather information and high-quality graphics is Weather Bank (5 Triad Center, Salt Lake City, UT 84108; 801-530-3181). This full-service weather information provider charges just $53 for their "Weatherbrief" software. Best of all, the connect charges are quite reasonable, ranging from 20¢ to 45¢ depending on your modem speed (up to 14,000 baud). You can even access WeatherBank through a toll-free number (for an additional 33¢ to 41¢ per minute). There are no minimums or monthly fees.

And what can you see? Weather Bank has color weather maps, satellite images, hurricane tracking maps, jet stream maps, lightning strike maps, and more. Of course, you can also access "raw" National Weather Service data such as current conditions, weather alerts (watches and warnings), aviation files (forecasts, winds aloft maps, etc.), and more. My pick for the best part of Weather Bank: high resolution state radar maps, updated every 30 minutes. These cost an additional $25 per month. Lower-resolution regional radar maps are included as part of the basic service, as are satellite pictures (updated every 30 minutes).

To access Weather Bank, you need an IBM PC or compatible computer that runs DOS 3.1 or higher, a hard disk drive, and an EGA or VGA graphics and color monitor. 512K of RAM must be free, and you also need a modem.

Let's say you don't have a fancy personal computer or high resolution monitor. Is there a weather computer service you can use to get weather data affordably?

The answer is yes. The Weather Network (568 Manzanita Ave. #1, Chico, CA 95926; 916-893-0308) is a small company that sells basic weather data at very reasonable rates. For a $35 registration fee, you get software that will run on any PC (XT or higher). Any monitor and version of DOS will work.

The rates to download data range from $1 to $1.50 per minute, depending on the speed of your modem (up to 9600 baud is supported). There is a $15 minimum charge per month. What makes the service affordable is the software (free with the $35 registration fee), which automates data retrieval for very quick results. Downloading your local forecast, three-to-five day outlook, and current conditions for your state would take about 15 to 20 seconds (about 50¢).

The Weather Network has only text-based weather data; sorry, no graphics or satellite pictures. The company takes National Weather Service data and loads it into its database programs to make it easier to use. You can get the data "raw"

143

or in an easy-to-read tabular-type format. Current conditions, forecasts, and upper-air data are available. Perhaps the most interesting information available on the Weather Network is the "forecasts discussions" between government meteorologists about weather outlooks, severe weather, and more.

Mark Welsh of the Weather Network explained that another advantage of their service is the company's small size. As a result, you can get personal service from their staff of meteorologists in case you have a question.

5. More Alternatives: Getting the Same Weather for Less Money

Looking at the above prices, I'm sure you suffered from a case of sticker shock. The fact is that there are ways of getting those fancy WSI and Accu-Weather images without mortgaging your home.

First and forecast, check out the on-line services such as CompuServe and America On-Line (see reviews Chapter 8). You'd be surprised at the number of offerings available on these easy-to-access networks. Best of all, the charges are more within the reach of the average weather buff.

Second, consider "re-sellers" who take expensive weather data and break it down into more affordable pieces. For example, Flight Data, Inc., offers a PC Weatherman for just $39.95 (call 800-451-3282 or 717-822-3236 or for more information). This affordable program accesses worldwide Accu-Weather satellite photos, regional radar maps, surface charts, lightning strike maps, and more . . . for a connect charge of just $1 per minute.

Weather buffs I've interviewed rave about PC Weatherman's radar maps, which have a "zoom and pan" feature to zero in on the details of a high resolution image. A unique "loop animation" feature lets you loop together graphic maps like satellite photos or NEXRAD images—the result is TV-quality graphics on your home computer. My only criticism: PC Weatherman is just for PCs; there is no Macintosh version yet.

If you like radar, you'll love Weather Watch by Say Soft (431 Roswell, Indianapolis, IN 46234; 317-271-3622). This $150 software program (for IBM-PC or compatibles) uses National Weather Service Aviation Weather data (easily downloaded from on-line services such as CompuServe) to create radar maps of the U.S. You can view the entire country or zoom in on your local area. Best of all, Weather Watch has county outlines and even highway (interstate, U.S., and state) information on its maps, making it easy to pinpoint storms. If you purchase Weather Watch, Say Soft will throw in a free CompuServe membership kit with $15 in free access time.

Satellite Photos without the Dish:
Tapping into Free Military Weather Broadcasts

9 hen I visited The Weather Channel to do research on the cable network for this book, I was amazed at the battery of electronic equipment in their "weather war room." Located behind the desk at which the on-camera meteorologist sits, this area was a weather buff paradise, with every meteorological tool you can imagine.

In an interview with the chief meteorologist, I asked which piece of weather information was the most important. If a power outage knocked out all the fancy computer systems and they could only have one piece of weather information, what would it be? The answer: satellite photos.

A good choice. Weather satellites provide a wealth of information that has greatly advanced weather forecasting. Satellites fill in the gaps between far-flung weather offices, showing the formation and movement of major weather systems before they wreak havoc on your Sunday picnic. But how can you get these valuable pictures without installing a giant satellite dish on your roof?

Surprisingly, there is a little-known and quite affordable way of getting the same satellite pictures you see on The Weather Channel: the U.S. military broadcasts them for free on shortwave radio. Until recently, only professional meteorologists and weather buffs with big budgets could access these broadcasts, called the "Military Shortwave Radio Weather Facsimile" network. The hardware and software to get weather "faxes" alone easily topped $1000—a major investment.

However, there is good news. Several companies have recently debuted packages of hardware and software to receive weather facsimile broadcasts that cost just $300. So, how does it work?

First, you need a shortwave radio receiver. Radio Shack sells these for about $200, and many manufacturers (see below) will sell you a complete package with a radio receiver and all other equipment you need. Once you have a way to receive shortwave broadcasts, you need a software package to convert the transmissions into pictures that you can view on a personal computer. This is where things get slightly more complicated. Basically, you need a "signal demodulator" that links the radio to your computer. Then you might have to add a "capture board" to your computer. As anyone knows who's done such a trick, this may involve significant brain damage.

Fortunately, there are several companies that offer complete packages with all the equipment and the know-how necessary to decode the shortwave transmissions with the minimum amount of pain. Software Systems Consulting

(714-498-5784), for example, sells a product called PC HF Fax for $100 that links the shortwave receiver to your computer and doesn't require a capture board.

Other companies that sell weather fax software and hardware include OFS Weather FAX (6404 Lakerest Ct., Raleigh, NC 27612; 919-847-4545; packages run $445 to $1690), Grove Enterprises (800-438-8155), Satellite Data Systems (507-931-4849), and Universal Radio (800-431-3939).

So, after you wade through all the techno-babble, what do you get? First of all, you can display satellite images of the U.S., broadcast 10 times daily. All of these companies offer software that lets you manipulate the data to your heart's content, looping pictures to create cloud movement and zooming in on a specific area. If that weren't enough, the weather facsimile network also twice daily broadcasts forecast charts for the next 24, 36, 48, and 84-hour period (that's the next three and half days).

Since the technical details of receiving weather facsimile broadcasts are beyond the scope of this book, I urge you to contact the above companies for more information. Most have brochures and newsletters that explain this process in more detail than you'll ever want to know. A good article on this subject entitled "Tuning in Free Weather Fax" appeared in the October/November 1992 issue of *Weatherwise* magazine. Check your library for a back issue or call Weatherwise at (202) 296-6267.

Talking to the Satellites:
Pulling down weather satellite pictures with Earth stations

10 Let's assume for a moment that money is no object. You're the Donald Trump of Weather Buffs. What's the ultimate weather toy? How about an "Earth station" system that receives pictures from just about any satellite that's floating above the planet.

We found a company that makes just a system: T-Ris Remote Imaging System by Telonics of Mesa, Arizona (932 E. Impala Ave., Mesa, AZ 85204; 602-892-4444). The cost? A mere $19,500. But, oh the places you'll go.

David Beaty of Telonics was gracious enough to send me some samples of the pictures that the T-Ris can receive.

Check out the picture of Hurricane Emily as it threatened the outer banks of North Carolina in August of 1993. You can actually see the 'details of the clouds swirling around a well-defined eye. The satellite pictures are so sharp, you can even see forest fires like the ones that swept Southern California in November 1993.

If all this is too rich, I did discover another company that makes satellite receiver systems that don't rival the cost of a fully-equipped Honda Accord. Vanguard Electronic Labs (196-23 Jamaica Ave., Hollis, NY 11423; 718-468-2720) has a whole catalog of products to "spy on the earth." Complete packages start at $700 and go up to $1050. You get an antenna, receiver, and all the hardware and software needed to display the pictures on an ordinary personal computer.

The Future

The weather information superhighway has come a long way in ten short years. Before The Weather Channel signed on in 1982, the only source for weather information was the local news—and what a poor source that was. Now, you can not only get weather 24 hours a day, but you can also download maps to a personal computer and use that same computer to plot and graph weather data from a home weather station.

If you think this is special, you haven't seen anything yet. In the next ten years, the amount of weather information (and the ease of getting it) will explode. Here's a peek at what's to come:

• More weather information available on-line. While there is already a wealth of weather on-line, there could be much more in the very near future. The Weather Channel is negotiating with a major on-line service to provide some of its custom maps on-line. Soon, you'll be able to download forecasts of rainfall totals or snowfall accumulations along with a plethora of other information. Look for more radar and satellite maps with incredibly high resolution. Easy-to-use software programs will let you pan and zoom the maps, as well as loop them together to create "images in motion."

• The high-tech weather radio. Right now, the weather radio that we wrote about in Chapter 6 is a pretty simple device: it receives the high-band frequency broadcasts from the National Weather Service (NWS). But the NWS is about to transform this device into an information powerhouse—the agency is mulling over transmitting a wealth of weather data (such as the hourly observations from the 600 weather reporting stations in the U.S. and Canada) on a companion frequency. Hence, your weather radio would be able receive free of charge reams of weather data that is now only available through expensive data services such as WSI and Accu-Weather.

How would this work? Well, you'd see inexpensive radio receivers and data converters on the market in no time. I have heard rumors that software companies are working on affordable programs that would let you use a personal computer to map, graph, and analyze this data. Best of all, this data would be FREE—no connect charges, monthly fees, and so on. Stay tuned.

• Doppler radar at 55 mph. With auto companies already experimenting with on-board maps and information screens in their cars, can weather information for the road be far behind? Just imagine cruising down the highway and suddenly noticing a big storm in front of you. Is there hail ahead? Will the heavy rain last 10 minutes or for the next 200 miles?

With a flick of a button, you call up the Doppler radar screen for your area. A geostationary location finder automatically calculates where your car is and displays your location as a blinking dot on the map. A quick look reveals the storm is a monster, but if you take the highway bypass exit, you'll avoid it all together.

A soon-to-be reality or mere fantasy? This technology may be closer than you think. As cellular phones go "digital," you may be able to get a lot more information on your car phone than just voice calls. A small television screen next to your car phone could display information on the weather, traffic, and more.

• Interactive weather "on demand." In Chapter 3, you got into a peek into the future of The Weather Channel. A spokesperson for the company said they're actively pursuing "weather on demand."

What the heck is that, you say? Well, imagine you're a hard-core skier, and the mountains have just got dumped on. With a flick of the remote, you'd be able to call up the skier forecast, the Michelin Driver's Report, and your local forecast. Instead of having to wade through Weather Channel segments that you don't want to see, you'd "call up" just the features you want. What you would see is a multi-media mix of graphics, maps, and "live-action" meteorologists giving you the latest reports.

For more information on this and other future projects The Weather Channel is experimenting with, turn back to Chapter 3.

So how can you stay on top of the latest developments in the weather information superhighway? *Weatherwise* Magazine is a good place to start; a one-year subscription is $32 (call 800-365-9753).

As soon as I'm aware of any new technologies, I'll include the information on the "update hotline" (1-900-988-7295) discussed in detail at the back of this book. Of course, I also hope to update future editions of this publication with information about the latest developments.

Feel free to call me at my office (303-442-8792) to ask a question or share your discoveries and suggestions.

In the Kitchen with the Earth: Recipes for Tasty Weather Phenomena

THINK OF THE EARTH AS A KITCHEN. And the weather as a series of old family recipes—a dash of moisture, a pinch of low pressure, a sprinkle of wind, and voilà! Weather à la the Earth.

In order to outsmart the weather using the tools in the last section, you have to know the basic ingredients in each weather recipe.

While this book isn't intended to be a textbook on meteorology (heavens, no), this chapter should give you a taste of some of our planet's most interesting weather phenomena.

So, join me as we take a look at culinary climatology. And no licking the beaters.

Thunderstorms

Thunderstorms come in several different flavors: the single cell or popcorn variety (you didn't think the food puns would end after the introduction, did you?), the squall line, and the "super-cell."

Popcorn thunderstorms get their name from the way they look on a satellite photo—a single, solitary cell of a storm that looks like a fluffy ball of popcorn. They can be as small as a mile or two wide at their base or as large as several dozen miles. Popcorn thunderstorms may produce just light rain, or on the other hand, develop into "super-cell" thunderstorms that can spin out a tornado. Squall lines usually precede a cold front and can be 250 or more miles in length. Thanks to their proximity to the clash between air masses, squall lines can drop heavy rain, hail, and other severe weather.

Ingredients for a Thunderstorm:
1. A supply of warm humid air
2. A trigger to give the aforementioned humid air an upward shove. The trigger could be one of several things: the heating of air

near the ground, a cold front that lifts the warm air ahead of it, mountains or hills that force the humid air to flow upward, or a sharp contrast between moist and dry air (West Texas often sees this phenomenon, called the Dry Line).

Steps:
1. Pre-heat the Earth's surface, which is covered with warm humid air.
2. Use one of the triggers listed above to help lift the air. These updrafts are crucial to the storm's formation.
3. Wring out water vapor in the air (condensation) to form droplets that make fluffy, white clouds (also known as cumulus clouds).
4. When the water drops grow too large, watch them drop back toward the Earth. This creates a downdraft and rain.
5. Simmer for a while. It's the mix of updrafts of warm humid air and downdrafts of rain the create the punch that thunderstorms need to create severe weather.
6. When the downdrafts outstrip the updrafts, the storm's heat source is removed, and it dies.

Makes 8 to 12 servings of rain—from as little as a trace to a "gully washer" with over one inch.

Serving suggestion: Sprinkle liberally on a city population that just washed their cars.

Variations:
 Some thunderstorms are meek, while others are killers. What makes the difference? Here are some variations on the above recipe that might produce a killer storm:
• Mix a fresh supply of warm humid air flowing on the ground with winds that increase with altitude. The result is unstable air that continues rising. Interaction with high altitude winds keeps the storm growing. These winds can tilt the storm, which keeps the rain from choking off the updrafts. The result: a storm that continues for several hours, creating dangerous lightning, high winds, hail, and possibly tornadoes.
• *Dry thunderstorms:* In the western U.S., popcorn-variety thunderstorms may have bases that are quite high. As a result, any rain evaporates before it hits the ground (this is called virga). "Dry" thunderstorms can still be dangerous because they may spit out frequent lightning and heavy downbursts of wind.
• *The Squall Line.* A strong line of thunderstorms that produces intense straight-line winds, hail, lightning, and possible tornadoes. Interestingly enough, these lines often form 100 miles or more in front of an advancing cold front (rather than right on the front itself).
• *Super-cell storms:* The Lion King of Thunderstorms. A supercell can be formed by itself or at the southwest end of a

squall line and can last for several hours. The recipe for super-cells involves a complex interplay of ingredients: the key is a flow of middle-level dry air. This creates a central updraft and can give the cloud the rotation necessary to spin out a tornado. The same updraft also keeps the storm alive for hours, and it can travel 100 miles or more.

Outsmarting Thunderstorms

The type of thunderstorm influences your ability to "outsmart" it. Popcorn thunderstorms can blow up right overhead with little or no advance warning. For these storms, consider investing in a lightning detector—the popcorn variety often produces lightning before any rain is detected on radar.

Squall lines are easier to forecast—check the weather maps for any strong cold fronts advancing on your area. The National Weather Service will issue a Severe Thunderstorm Warning for areas that might see a significant outbreak of storms—make sure your weather radio is set on alert to hear the latest watches and warnings. You might want to keep the radio at a central location (a kitchen, for example) as well as additional radios in the car, workshop, etc.

No matter what the flavor, all thunderstorms need a supply of humid air. How can you tell how much humidity is around? Check out The Weather Channel—in the summer, they display a map of dew points, indicating how "juicy" the atmosphere is (and, hence, how primed for thunderstorm development). Dew points above 60° in the summer indicate that conditions might be conducive to thunderstorm formation.

When a storm is heading your way, you can use your home weather station to watch the rising humidity and falling barometer. Using the sources in Chapter 8 and 9, you might be able to download a radar picture on your home computer. When trying to predict the direction a thunderstorm is heading, remember that while the entire line of thunderstorms may be moving in one direction, *individual* cells may be moving in another direction altogether. It's not uncommon, for example, to see a squall line advancing to the southeast at 15 mph, while individual cells are moving northeast along the line. As a result, you'll need to look "down the line" to some degree to see which part of the line might hit your location.

Tornadoes

 If a picture is worth a thousand words, then a piece of live-action video must be worth ten times more. While I wish there could be a CD-ROM companion to this book with tornado footage, I suppose you'll have to be satisfied with the twister footage that's shown occasionally on The Weather Channel.

Meanwhile, let's take a look at the recipe for one of the most terrifying weather phenomena on the planet.

Ingredients:
1. *One super-cell thunderstorm*
2. *A supply of cool, dry air flowing into the storm at middle altitudes*

Steps: Interestingly enough, scientists are still trying to figure out how tornadoes actually form. Sure, we know that when the above ingredients are mixed, the result may be a rotation in the cloud—this may form a wall cloud, from which a funnel may appear. Yet watching tornado footage makes one aware of how complex tornadoes are. Some have smaller "satellite" funnels that revolve around a main twister. The sheer difference in size and scale is amazing as well. The following "measuring tool" sizes up the differences in tornadoes:

Measuring tool: Fujita Wind Damage Scale—classifies tornadoes by wind speed

Type	Wind Speed	Damage
F-0	*up to 72 mph*	Light Damage
F-1	*73-112 mph*	Moderate
F-2	*113-157 mph*	Considerable
F-3	*158-206 mph*	Severe
F-4	*207-260 mph*	Devastating
F-5	*above 261 mph*	Incredible

Serving Suggestion: Tornadoes and automobiles don't mix. Many people who try to flee a twister are killed—in fact, most of the fatalities in tornado outbreaks are found in the twisted wreckage of cars tossed about in the wind. A better bet: take shelter in a building or, as a last resort, in a ditch.

Variations:
One little-known, less-publicized fact about tornadoes is that they have funnel "cousins." Here's a quick overview of the variations:

• *Gustnadoes.* These are formed in the outflow of air from a major thunderstorm. These whirlwinds have top speeds that rarely exceed 150 mph and last less than five minutes. Yet, the vortex can do some damage—especially if your mobile home happens to be in the path of one.
• *Landspouts.* Landspouts are grown-up versions of gustnadoes. Formed in much the same way, landspouts rarely exceed F2 (157 mph) but can last much longer than gustna-

154

does. These funnels are examples of "non super-cell tornadoes," vortices that are attached to clouds that you wouldn't think would spawn a tornado. A fascinating article on and picture of a 1988 landspout in Australia appeared in *Weatherwise* magazine (June/July 1994, page 37). Just like a new species of plant, landspouts have only been recently identified and studied by scientists. This type of funnel is common in the plains of Oklahoma and Colorado.

• *Dust devils.* Anyone who's traveled the deserts or high plains can probably tell you the story of a huge funnel of dust they saw spinning in a farmer's field. Dust devils can reach a hundred or more feet in height and have winds in the F0 (under 72 mph) range. Most are caused by the sun heating the unstable air near the ground—a whirlwind of dust may be caused by a tractor, passing car, or other disturbance.

Outsmarting Tornadoes:

Let's be realistic: it's hard to outfox a weather phenomenon that can strike at any time with little warning and that can pack winds in excess of 200 mph. Yet there are a couple tips to try to predict killer tornadoes.

Since most tornadoes are born in super-cell thunderstorms, watch the sky when the National Weather Service has issued a Tornado Watch for your area. A weather radio is a good investment, as is keeping your eye on the local weather radar. If you can download NEXRAD weather radar pictures, look for wind shear near a strong thunderstorm—winds flowing into the storm from two different directions show up as two distinct colors on the radar image.

A scanner is a good investment if you live in tornado country. You can pick up the chatter of weather spotters watching for wall cloud development, a precursor to funnels. Scanners also will allow you to be privy to damage reports for nearby towns hit by the storm first—the police radios will crackle with details about any destruction and calls for help, giving you an idea of an approaching storm's strength.

If a tornado is spotted, the National Weather Service will issue a tornado warning, activating special weather alert radios with the information. Unfortunately, the weather service doesn't bat one-thousand on this one—a tornado may strike without any warning.

Most people who have been through this experience describe the roaring winds of the funnel as the sound made by an oncoming train. If you hear this, you have very little time to react, so it's important to know some basic tornado safety tips.

How do tornadoes destroy a house? Interestingly enough, it has nothing to do with the myth that low pressure in the funnel implodes the structure. This led to the faulty advice to open the windows before a tornado hits in order to "equalize" the pressure in the home. Actually, with the windows open, your house is more likely to be damaged (and you injured) by a piece of debris, such as a two-by-four flying at 95 mph through your living room.

The best advice: forget the windows and head to a "safe area" in the house. In order to know what area is most safe, you should know that most tornadoes destroy a home by first knocking down the wall that is first in its way. Then the roof is lifted off and the remaining walls fall outward like a house of cards that has had its top removed. Therefore, a good place to be is in an interior room on the lowest level of the house (such as a bathroom—the plumbing helps stiffen its structure). Seeking shelter in a room with windows or near an exterior wall might be a fatal mistake.

A common piece of tornado safety advice is to take shelter in a basement. Yet, this seems silly since most tornadoes strike places like Texas and Oklahoma—states that have very few homes with basements.

Of course, tornadoes can strike in just about any place in the U.S.—in any month and at any time of the day. That's what makes tornadoes so fascinating and scary—no other weather phenomenon can be so ferocious and so unpredictable at the same time.

Heat Waves

 I'll never forget being a weather junkie while growing up in Texas—especially during the 1980 Summer from Hell. Now, as anyone who's dared to venture to the Lone Star State in August knows, it's always sizzling during the summer. Add humidity, and you have a state that could be reclassified as the world's largest sauna.

Yet the summer of 1980 redefined the word "hot," especially in the Dallas/Ft. Worth metroplex. The cities recorded 42 days of high temperatures above 100°F. On June 26 and 27, the mercury topped out at a whopping 113°F. I remember going outside that day in Dallas and standing on the sidewalk while waves of superheated air washed over me. I now know what it feels like to be a slab of beef at a barbecue.

Wichita Falls did Big D one better and registered 117°F. Factor in the humidity and it probably felt like Venus. Of course, Texas wasn't the only state that suffered during the summer of 1980. With the exception of New England, every state (that's nearly all 48 in the continental U.S.) recorded highs above 100°F during July 1980.

With all this personal experience, I humbly offer my recipe for a heat wave.

Ingredients:
1. One large dome of high pressure
2. A weak upper-air pattern
3. A polar jet stream that carries the storm track well into Canada

156

Steps:

1. Take the high pressure and bake the ground. The cloudless sky keeps the heat on, building day after day.

2. Feed the warmer air into the high-pressure system, making it stronger. At the same time, the stronger high pressure blocks any humid air from the Gulf of Mexico from forming relief-producing thunderstorms.

3. During the summer, the extended daylight hours continue the baking process.

4. How do you break a major heat wave? One way is a major hurricane. In August 1980, Hurricane Allen (no relation to the author) slammed into the lower Texas coast, just north of Brownsville. It took this super-low pressure (the hurricane had a central pressure of just 990 millibars and winds of 195 miles per hour) to knock out the super-high pressure that was responsible for the summer of 1980.

Serving Suggestion: Serve sunny-side up on a major holiday when every air-conditioning repair company is on vacation.

Variations:

One of the fascinating things about heat waves is that they often create a Dr. Jeckyl/Mr. Hyde weather scenario. One part of the country bakes, while another basks in cool, dry Canadian air. Why? Because the high-pressure system, while huge in size, usually only covers one part of the country. At the same time, the jet stream (or storm track, as it is called), pulls cooler air from Canada into the other part of the U.S. Depending on which side of the jet stream you're on (the distance of just a few hundred miles), you could be baking in the heat for weeks.

Outsmarting the Heat Wave:

Heat waves may be brutal, but they do have one good aspect—you can often see one coming. Heat usually builds in one area (like the Desert Southwest) and spreads to other parts of the country on southwesterly winds. You can see this process starting by monitoring high temperatures (easily downloaded from a computer on-line service such as those mentioned in Chapter 8).

While you'd expect a place like Phoenix to be above 100°F in June, you know you're in trouble in Boise when super-heated air makes it as far north as Salt Lake City. Similarly, heat that builds in the Great Plains may soon travel east. Once it hits the century mark in Kansas City, folks in Chicago know they're next in line.

The key to forecasting a heat wave may be watching the upper-air jet stream pattern. The Weather Channel often includes a jet stream forecast map in its report at the top and bottom of the hour. Look for a big ridge of high pressure in the upper atmosphere to send the storm track way into

Canada. To the north of the jet stream, the air may be pleasantly cool. On the other side, it's sweat city.

Winds are another key element to heat waves—southwesterly winds are often needed to suck super-hot air out of the deserts of Mexico and spread it toward more temperate areas like New England. During some times of the year, a surface high-pressure system in the Atlantic can provide those warm southwest winds. When forecasting a heat wave, it helps to be able to recognize in what position a high-pressure system must be in order to result in favorable southwesterly winds.

If you have a home weather station, you can predict a warm day by discovering a pattern to your area's overnight lows in the summer. At my house at an altitude of 7400 feet in the foothills of Colorado, the temperatures at night usually fall into the 50s and low 60s. But when the low temperature barely brushes the 70° mark and the mercury by 10 AM is in the low 80s, we know we're in for a warm one.

Blizzards

 I have to admit to a strange weather affliction: I love blizzards. As long as I'm not caught driving in one, I should add.

Blizzards are one of my favorite weather phenomenon because of their sheer size and impact. No other weather event has the potential to paralyze a city like a 12-inch blizzard. While thunderstorms can wreak damage over a wide area, only a blizzard can shut down the entire eastern seaboard for days.

The combination of heavy snow and high winds can be triggered by very different conditions, depending on which part of the U.S. you live. In California, a major winter storm can be caused by a Pacific storm that slams into the Sierra Nevada. The position of a low-pressure system that forms on the eastern slopes of the Rockies can bring two-foot snows to Cheyenne, while Denver is basking in sunshine and warm temperatures. On the East Coast, a slight shift in the track of a "Nor'Easter" storm can blast certain cities while leaving others unscathed. In the South, winter storms add a particularly wicked ingredient: ice. Just a thin coating of ice can send cities like Dallas into a tizzy.

So, that's what I love about blizzards. Here's my recipe for a decent blizzard that would affect much of the U.S.:

Ingredients:
1. A dome of arctic air from Canada
2. Warm, humid air from the Gulf of Mexico
3. A deep low-pressure system

Steps:
1. Take the low-pressure system and slam it into the West Coast, dumping heavy wet snow in the Sierra Nevada or Cascades of Washington and Oregon.
2. Compact the storm as it travels over the Rocky Mountains. Like a spinning ice skater who pulls her arms in, the storm rotation speeds up as it gets more compact.
3. Pop the storm out on the eastern slopes of the Rockies. Once it hits the plains, the storm will spread out and tap the aforementioned moisture from the Gulf of Mexico.
4. Compact the moisture from the storm against the higher elevations of the Front Range. This can produce significant snow from eastern New Mexico to Montana, depending on where the low pressure forms.
5. As the storm crosses the plains, trail a cold front that sparks thunderstorms across the southern U.S. As the moisture wraps into the colder Canadian air behind the front, precipitation may turn into rain, sleet, and finally snow.
6. Slam the storm into the Appalachians. The storm's center may dissipate, only to form a secondary low-pressure center off the East Coast. Take this second low and strafe the East Coast before heading out to the Atlantic, dumping snow on Maine and northern New England.

Serving Suggestion: For maximum television coverage, serve over a major media market like, say, New York City. Mix with a dash of freezing rain and a major travel holiday to really create havoc.

Variations:
Depending on the topography and terrain, blizzards and major winter storms can appear in one of several variations. Here are some common options:

• *Lake Effect Snow.* Instead of low pressure, substitute a strong high-pressure system based in Canada. Flowing clockwise from the high pressure, send Arctic cold winds howling over the waters of the Great Lakes. As the wind travels over the warmer water, it picks up moisture that immediately cools and forms clouds. These clouds can dump huge amounts of snow on the cities right along the lake's shoreline.
• *Upslope Snow.* Live in Boulder, Colorado, for any length of time and you'll become an expert on upslope snow. Basically, the recipe for upslope snow involves easterly winds that push air up the slope of the foothills of the Rocky Mountains. As the air rises, it cools, forms clouds, and eventually, large amounts of snow. Unlike other forms of snow storms, upslope conditions can be caused by *either* high- or low-pressure systems. Basically, any system that sends easterly winds toward Boulder (for example, a low pressure in northeastern New Mexico or a high-pressure system over

Nebraska) can create conditions for a major snow.

One of the ironies of snowstorms in the west is what we locals call the "Denver Bronco Effect." Whenever the Broncos play "Monday Night Football" at home and, at the same time, a major snowstorm hits the city, the state's ski resorts are flooded with reservations. The irony is that the same storm that socks Denver with the heavy snow seen on TV usually brings little or no snow to the Rockies (where the ski resorts are located, about two or three hours to the west). Why? Upslope snows are a very "localized" phenomenon— while easterly winds bring snow to the Front Range, the clouds and moisture don't travel over the Continental Divide, leaving the ski resorts high and dry.

• *Ice Storms.* A good old-fashioned ice storm in Texas is an interesting variation on the winter storm recipe. Unlike the upslope and lake effect snow conditions that occur common- ly throughout the winter, an ice storm is caused by a rare convergence of several ingredients. First, you need a strong cold front with Arctic air to sweep across Texas. Because of the state's southern latitude, by the time a cold front reaches the Gulf of Mexico, the pool of cold air is quite shallow (per- haps just 1000 or so feet deep). To create an ice storm, you need an upper-level low-pressure system to skirt the southern border of the U.S and clash with cold air. When the low- pressure system reaches Texas, the warmer air aloft means the precipitation will start out as rain. As it falls through the colder air, it may turn to sleet or freezing rain. The critical ingredient is the air temperature at the ground—if it's near 32°F, the precipitation may stay in a liquid form until it comes in contact with the ground (or roads or electrical wires, etc.), at which point it forms ice.

Outsmarting the Blizzard:

Winter storms probably produce the most "false warn- ings" of all the severe weather phenomena—predicting the exact path of a low-pressure system is tricky at best. Nonetheless, I've found that the National Weather Service can successfully predict a major winter storm at least one or two days in advance. The "Storm of the Century" in 1993 that blasted the East Coast with up to 50 inches of snow was predicted more than a day in advance.

Once again, the inexpensive weather radio is a good first line of defense—you'll hear the National Weather Service issue a winter storm watch and possible warning as condi- tions warrant. While I like the advance notice, I've noticed the rocket scientists at the weather bureau seem somewhat gun-shy when it comes to predicting snowfall accumula- tions—it seems they wait for the snow to start to fall before they begin guessing the totals.

A better source for snowfall projections is The Weather Channel—their maps often show accumulation estimates hours before the National Weather Service announces its prediction.

I've found a scanner to be an invaluable tool in monitoring the progress of a blizzard. You can listen to state patrol officers talking about closing a certain highway because of icing. The latest accidents clue you into obvious trouble spots like overpasses and bridges (which always ice up first because they are exposed surfaces that don't have the benefit of heating from the ground). As you get more familiar with your area's topography, you'll recognize a particular location that ices first, such as a hilly area or highway overpass—a clue to the storm's progress.

Unfortunately, weather radar isn't much help in predicting how much snow will fall. Most radars have trouble distinguishing between different levels of snowfall and sometimes light snow is invisible to the radar altogether. The new NEXRAD radars should help fix this problem to some degree, since their increased resolution can "see" snow (and distinguish between different rates of precipitation) better than previous generations of radar.

Hurricanes

 There's nothing like a major hurricane to ruin your whole day. These awesome storms are truly unique in the weather kitchen—how many other storms can last for a week, affect areas hundreds of miles long, and, best of all, have their own names? Here's a recipe for your basic, garden-variety hurricane:

Ingredients:
1. Wide expanses of warm ocean water. This water needs to be above 80°F and at least 250 feet deep.
2. Warm, humid air up to about 18,000 feet—this feeds the storm.
3. Unstable air, which is the "yeast" of hurricanes, keeping the storm going. In the Atlantic, this humid air needs to be about 5° to 15° latitude (north of the equator) to provide just the right conditions for storm development.
4. Upper-air winds blowing in the same direction as winds near the surface. If this doesn't happen, the upper-air winds will shear off the tops of the storm clouds, bringing a premature end to the storm.
5. Upper-atmosphere high pressure. This helps pump air away at the top of the storm.

Steps:
1. Mix the humid air with a trigger, perhaps a disturbance moving off the West African coastline.
2. Form a cluster of thunderstorms (convection) that blossom in the humid, unstable air.

3. Add favorable upper-air winds to create a "tropical cyclone," the formal name of these storms. Here's the measuring tool for tropical storms, known as the Saffir-Simpson Scale:

NAME	DAMAGE	WINDS (in mph)	BAROMETRIC PRESSURE (at storm's center, in inches)	STORM SURGE (in feet)
Tropical Depression	Little	Less than 39 mph	Above 29 inches	None
Tropical Storm	Minor flooding	39 to 74 mph	Above 29 inches	Very minor
Category 1 Hurricane	Minimal damage	75 to 95 mph	Above 28.94 inches	4 to 5 feet
Category 2 Hurricane	Moderate	96 to 110 mph	28.50 to 28.91 inches	6 to 8 feet
Category 3 Hurricane	Extensive	111 to 130 mph	27.91 to 28.47 inches	9 to 12 feet
Category 4 Hurricane	Extreme	131 to 155 mph	27.17 to 27.88 inches	13 to 18 feet
Category 5 Hurricane	Catastrophic	Above 155 mph	Less than 27.17 inches	More than 18 feet

Serving Suggestion: Combine a major hurricane with lax building regulations and poof! Major disaster! Make sure builders quickly construct houses with inadequate bracing, cheap building materials, and poorly installed roofing to ensure maximum damage.

Outsmarting Hurricanes:

Over 400 hurricanes and tropical storms have hit the Atlantic and Gulf of Mexico coasts since 1870. Despite all this experience, scientists are still at a loss in trying to answer the key question: when will the next one hit and where?

While we don't know exactly how many hurricanes will form each year, once one does form, there is at least some advance notice before it makes landfall. This can be a day or two, or even up to a week in some cases.

How can you outsmart a hurricane? A good place to start is the "Tropical Update," a regular feature of The Weather Channel during the hurricane season (which officially runs from June 1 to November 30). Airing between 10 to 15 minutes before the top of the hour, this valuable report is your first early warning of tropical development and, incidentally, a nice lesson on hurricane formation and behavior.

Once a tropical storm forms, you can use one of several affordable programs to track the storm's progress on your home computer. One of the best is Hurrtrack, published by PC

Weather Products (PO Box 72723, Marietta, GA 30007-2723; 800-242-4775 or 404-953-3506). The software comes in two editions: regular Hurrtrack ($49) and the professional version of Hurrtrack ($149). You need an IBM PC or compatible, VGA graphics card and monitor, 640K RAM, and a two-megabyte hard disk (seven megabytes for the professional version).

This amazing program includes seventeen detailed tracking maps with over 200 city locations, graphical representations of the eye, destructive eyewall, tropical storm force, and hurricane force winds. Best of all, you also get a forecast feature that uses a 100-year storm database to compute a statistical forecast track.

The professional version has ten additional "highly-detailed" tracking charts with counties, roadways, and more. You can also enter and display the National Hurricane Center *forecast* data for the storm, which shows meteorologists' best guess for possible strike areas. The professional version also has the ability to modify the storm's eye characteristics (eye diameter and eyewall size).

In trying to predict most hurricanes, recognize that many storm tracks are influenced by a regular summer phenomenon called the Bermuda High, a dome of high pressure that is usually centered over (surprise) Bermuda. When the high pressure is small, hurricanes often veer away from the East Coast. However, a large Bermuda high may steer a hurricane right into the populated areas of the coastline. The trickiest hurricanes to forecast are the ones that form when upper-air currents are quite weak—these hurricanes can wobble about aimlessly, stopping at one point, then starting again, only to change course several more times.

How can you make sure your home survives a hurricane? Hurricane Andrew provided several pointed lessons about which construction techniques work and which don't. A fascinating analysis of "what went wrong" appeared in the *Miami Herald* (December 20, 1992). Article after article dissected the storm, pointing out "builders' shortcuts to disaster." The stories tell how the areas hit hardest by Andrew were not in the path of the storm's strongest winds. In fact, homes that were built most recently (no matter where they were located) fared worse than older homes. Why? Sloppy construction, watered down building regulations, and inadequate design were much more common in homes built after 1980, not before. As one article pointed out, "all too much damage was avoidable. Builders at these developments seemed to underestimate the risk of a 'Big One.' Homeowners paid the price."

Do you have a recipe for a local weather phenomenon you'd like to share with our readers? Call me at 303-442-8792 or write to the address at the back of this book.

Part Four

the best gifts for the weather junkie

The Top 10 Gifts & More for Weather Junkies

I T'S GIFT-GIVING TIME AGAIN, and you're drawing a blank—just what do you give a weather junkie? It's not like you can go to the mall and pop into the "weather store" to find the perfect gift—many of the best weather gifts are only available thorough obscure catalogs that cater to weather enthusiasts. As a result, I'd like to humbly present my list of the best gifts for weather junkies.

1. BRIGHTER DAY SKY UMBRELLA

Sure, it looks like a normal black umbrella—but wait! Open it up to find a colorful blue sky with fluffy clouds inside. If you're tired of the rain, this might be just the trick. Cost: $89 (Item #3A137). Available from the Museum Collections catalog (800-442-2460).

2. ALASKA WEATHER CALENDAR

Sure, you've seen a zillion nature calendars at the mall. But you haven't experienced the ultimate in weather calendars until you've checked out the Alaska Weather Calendar, designed by Jim Green, an Anchorage meteorologist and photographer.

Stunning photographs highlight each page, along with a plethora of weather facts, maps, and graphics. Among the more interesting sections: commemorations of major weather events, such as the August 15, 1967, flood when the Tanan River crested several feet over its banks. Six people died, and there was $100 million in damage in Fairbanks and Nenana. As if all this isn't enough, the Alaska Weather Calendar even has sun-finder charts, windchill graphs, and a complete map of Alaska with average temperature graphs for specific climate zones. If you've ever wondered what the weather is like in the fiftieth state, this is *the* gift. Cost: $10 plus $2 shipping. Williwaw Publishing Company (800-490-4950).

3. "BARBEQUE METEOROLOGY" SWEATSHIRTS AND T-SHIRT

"It's an immutable law of nature: The position of the chef determines the direction of the wind." Any barbeque enthusiast can relate to the cartoon on these shirts showing the chef choking in masses of smoke no matter which direction he cooks from. The perfect gift for the meteorologically inclined cook. Cost: $28 for the sweatshirt and $18-20 for the T-shirt. From the Wireless catalog (Item #35401 or 35400, 800-669-9999).

4. KANSAS STORMS: DESTRUCTION, TRAGEDY & RECOVERY 1991

1991 was one of the worst years for killer tornadoes in Kansas—over 115 twisters struck the state in the first six months alone. *Kansas Storms* is a fascinating chronicle of the people, events, and weather that marked that violent year.

What separates this book from other titles on the subject are the fascinating insights into each storm by the people who lived through it. These personal retrospectives are haunting. You'll read about the doctor who raced home to get his wheelchair-bound wife out of harm's way and the television reporter who ducked beneath an interstate underpass, narrowly being missed by a fast-moving vortex. Seventeen pages of color photos and hundreds of black and white shots capture some amazing moments, from cars wrapped around trees, to houses that were torn to shreds in a matter of minutes. This should be on the bookshelf of any weather buff. Cost: $14.95, paperback. Published by Hearth Publishing, 135 N. Main, Hillsboro, KS 67063. Call (316) 947-3966 to order with a credit card.

5. A SUBSCRIPTION TO *WEATHERWISE* MAGAZINE

What *Rolling Stone* is to music and *Newsweek* is to news, *Weatherwise* is to weather.

It's not just filled with scientific research on meteorology—the articles focus on current topics in the weather, from major blizzards to catastrophic floods. Regular sections cover the latest weather software, videos, and books. My favorite is the yearly wrap-up of the weather—you get a blow-by-blow account of the previous year's weather, told with pictures, maps, and graphics.

Weatherwise comes out bi-monthly and is a must for any weather junkie. Cost: $32 for one year (six issues). Published by Heldref Publications, 1319 Eighteenth St., NW, Washington, D.C. 20036. Call (800) 365-9753 to order.

6. ONE-FOR-THE-ROAD THERMOMETER

You're driving along in a rainstorm when suddenly the precipitation turns to sleet and then to snow. Will the road ice over? You can answer this question in an instant with a car thermometer that displays the outside temperature. I like using the "Digital Indoor/Outdoor Thermometer and Clock" from the Wind & Weather catalog for this purpose. The outdoor sensor has a self-sticking back that mounts on the driver's side mirror; you can snake the wire out the driver's side door.

And here's the best part: the display has easily readable 3/4-inch numbers. Flip a switch and change from outdoor to indoor temperature; hit the bar and you get the time. The temperature range is -58°F to 158°F—perfect for even the harshest climates. Cost: $24 plus $1.50 shipping. Item #IN-DTC1 from the Wind & Weather catalog (800-922-9463).

7. THE PACESETTER SIX WEATHER STATION

This is the most elegant weather station on the market, with its sleek black panel that displays an amazing array of weather information. The Pacesetter Six Weather Station from Oregon Scientific gives you indoor temperature, humidity, barometric pressure, clock, and date. And here's the neat part: a "forecast" window shows conditions for the next 24 hours (clouds, sunshine, partly sunny, etc.). A bar graph also tracks the barometric pressure trend over the past day.

Of course, it isn't cheap, retailing for $140 in many catalogs, such as Hammacher Schlemmer (800-543-3366). As this book went to press, I noticed that a second, more affordable version has just debuted. This model has an indoor/outdoor thermometer, barometer with trend arrow, and weather forecast window. I've only seen the new model in the Safety Zone catalog ($89.95; Item #855080, 800-999-3030 or 717-633-3370), but I assume it will be widely available soon.

8. CASIO OUTDOORSMAN WATCH

Looking for a way to have your weather station with you no matter where you go? Casio has made it possible with their Outdoorsman Watch. You get a digital compass, altimeter, barometer, thermometer, and stopwatch. Casio has even built in a graph of the barometric pressure—perfect for forecasting on the go. Cost: $199. Available from the Sharper Image catalog (Item #C0305, 800-344-4444).

9. GALILEO THERMOMETER

Yes, it was back in the good ol' 1600s when Galileo invented the "slow thermometer" using the principle that the density of a liquid changes with temperature. Hence, if you take a long tube of glass, fill it with water, and throw in several globes containing a dense liquid, you can determine the temperature. As the temperature rises, the glass globes descend one by one, since the liquid they're suspended in is less dense and not able to support their weight. (There will be a test on this after the chapter.)

Even if you don't understand how it works, the Galileo

Thermometer will make for an interesting conversation piece. Prices range from $195 to $255, depending on the size you choose. You even get a choice in colors—burgundy and dark blue! Available from the Wind & Weather catalog (Item #IN-TL17 or IN-TL24, 800-922-9463).

10. GIFT CERTIFICATE TO THE NATURE COMPANY

Just drop me off at the Nature Company and you don't have to come back for hours—I love this place. They've got everything any weather buff could want, from instruments to books, almanacs to telescopes. (Okay, that last item isn't exactly used to watch the weather, but the Nature Company does have a cool collection of telescopes.)

If you can't decide what to buy, the Nature Company has gift certificates, available in any denomination. You can use them at any of the stores or when ordering from their catalog (see description below). Call (800) 227-1114 or (800) 607-7888 for more information.

Weather Catalogs

I've scoured the ends of the earth for the best catalogs of weather gadgets. Here they are:

◆ **American Weather Enterprises**
What They Sell: Weather instruments and more
To Order Call: (215) 565-1232
Shopping Hours: 9 AM to 7 PM, Monday through Friday; 9 AM to 1 PM Saturdays; also have an answering machine at other times
Or write to: American Weather Enterprises, PO Box 1383, Media, PA 19063
Credit Cards Accepted: MC, VISA

◆ **Bass Pro Shops Marine Catalog**
What They Sell: Yes, it's primarily a catalog for bass fishers, but they do have an interesting collection of thermometers for boats and other outdoor uses.
To Order, Call: (800) 227-7776 or (417) 881-3567; TDD (800) 442-5788; Fax (417) 887-2531
Shopping Hours: 24 hours a day, seven days a week.
Or write to: Bass Pro Shops, 1935 S. Campbell, Springfield, MO 65898
Credit Cards Accepted: MC, VISA, AMEX, Discover

◆ **Edmund Scientific Co.**
What They Sell: The granddaddy of science catalogs is a must for anyone who likes meteorology, astronomy, and anything science related. Their weather section includes instruments, books, experiments, and more.
To Order, Call: (609) 547-8880; Fax (609) 573-6295
Shopping Hours: 8 AM to 8 PM, Monday through Friday; 9 AM to 4 PM, Saturdays
Or write to: Edmund Scientific Co., 101 E. Gloucester Pike, Barrington, NJ 08007
Credit Cards Accepted: MC, VISA, AMEX, Discover, Optima, Diners Club
Retail Store: Call (609) 573-6241 for directions if you're in the Philadelphia area.

◆ **Hammacher Schlemmer**
What They Sell: While Hammacher Schlemmer mostly offers toys for yuppies, this catalog has some interesting weather gadgets from time to time.
To Order, Call: (800) 543-3366 or (800) 233-4800; Fax (800) 444-4020
Shopping Hours: 24 hours a day, seven days a week
Or write to: Hammacher Schlemmer, 9180 Le Saint Dr., Fairfield, OH 45014
Credit Cards Accepted: MC, VISA, AMEX, Discover, Diners Club, Carte Blanche

◆ **Heartland America**

What They Sell: You'll find scanners, weather stations, and thermometers in this discount catalog. Did I mention that they sell at discount prices?

To Order, Call: (800) 229-2901; Fax (800) 943-4096

Shopping Hours: 8 AM to 5 PM, Monday through Friday

Or write to: Heartland America, 6978 Shady Oak Rd., Eden Prairie, MN 55344

Credit Cards Accepted: MC, VISA, AMEX, Discover

◆ **Into the Wind**

What They Sell: Looking for a funky wind meter? This catalog has several.

To Order, Call: (800) 541-0314 or (303) 449-5356; Fax (303) 449-7315

Shopping Hours: 9 AM to 6 PM, Monday through Friday; 9 AM to 4 PM Saturdays

Or write to: Into the Wind, 1408 Pearl St., Boulder, CO 80302

Credit Cards Accepted: MC, VISA, AMEX, Discover

Retail Store: Right here in my own hometown of Boulder, Colorado, on the beautiful Pearl Street Mall

◆ **KLOCKIT**

What They Sell: Besides clocks, KLOCKIT sells components for do-it-yourself weather stations.

To Order, Call: (800) 556-2548 or (414) 248-9899

Shopping Hours: 7 AM to 8 PM, Monday through Friday; 8 AM to 4 PM Saturdays

Or write to: KLOCKIT, PO Box 636, Lake Geneva, WI 53147

Credit Cards Accepted: MC, VISA, Discover

◆ **The Nature Company**

What They Sell: The Nature Company produces a beautifully illustrated and photographed catalog featuring a collection of "tools for the naturalist." Some items are available in their stores, while others are available only through the catalog. In the "Curiosity" section, I noticed the Davis Weather Monitor ($395) with all its accessories. Another interesting section is the new "multimedia" area, with CD-ROMs on astronomy and dinosaurs, among other subjects. This wonderful catalog should be on the coffee table of any weather enthusiast.

To Order, Call: (800) 227-1114 or (800) 607-7888; Fax (606) 342-5630

Shopping Hours: 24 hours a day

Or Write To: 750 Hearst Ave., Berkeley, CA 94710

Credit Cards Accepted: MC, VISA, Discover or American Express

Retail Stores: Call the numbers above for a listing of their stores

◆ **Northeast Discount Weather Catalog**
What They Sell: Good quality weather instruments at discount prices
To Order, Call: (800) 325-0360
Shopping Hours: 9 AM to 5 PM, Monday through Saturday
Or Write to: Northeast Discount Weather Catalog/Viking Instruments, 532 Pond St., S. Weymouth, MA 02190
Credit Cards Accepted: Discover, VISA, MC

◆ **The Safety Zone**
What They Sell: Just about any safety product you can think of, including weather radios, thermometers, and more
To Order, Call: (800) 999-3030 or (717) 633-3370; Fax (800) 338-1635
Shopping Hours: 24 hours a day, seven days a week
Or write to: The Safety Zone, PO Box 692, White Plains, NY 10604
Credit Cards Accepted: MC, VISA, AMEX, Discover

◆ **The Sharper Image**
What They Sell: The Casio Outdoorsman watch and Pacesetter Six weather station (see descriptions above)
To Order, Call: (800) 344-4444 (fax orders, press 3) or (800) 344-5555
Shopping Hours: 24 hours a day, seven days a week
Or write to: The Sharper Image, 650 Davis St., San Francisco, CA 94111
Credit Cards Accepted: MC, VISA, AMEX, Discover, Diner's Club

◆ **Sporty's Pilot Shop**
What They Sell: Everything a pilot might need—including some interesting weather gadgets, books, and videos
To Order, Call: (800) LIFTOFF (543-8633); Fax (513) 732-6560
Shopping Hours: 10 AM to 6 PM, Monday through Friday
Or write to: Sporty's Pilot Shop, Clermont County Airport, Batavia, OH 45103
Credit Cards Accepted: MC, VISA, Discover

◆ **Sporty's Preferred Living**
What They Sell: Rain gauges, weather radios, outdoor thermometers, and more
To Order, Call: (800) 543-8633 or (513) 732-2411; Fax (513) 732-6560
Shopping Hours: 10 AM to 6 PM, Monday through Friday
Or write to: Preferred Living, Clermont County Airport, Batavia, OH 45103
Credit Cards Accepted: MC, VISA, Discover

◆ **Weatherama Weather Instruments**
What They Sell: Weather instruments at decent prices
To Order, Call: (800) 848-4912, ext. 3187, or (612) 432-4315;
Fax (612) 432-9754
Shopping Hours: 8 AM to 6 PM central time, 24 hour voice mail
Or write to: Weatherama, 7395 162nd St. West, Rosemount,
MN 55068
Credit Cards Accepted: MC, VISA

◆ **Robert White Weather Instruments**
What They Sell: Professional and technical instruments
To Order, Call: (800) 992-3045; Fax (617) 742-4684
Shopping Hours: 9 AM to 5 PM, Monday through Friday
Or write to: Robert White Weather Instruments, 34
Commercial Warf, Boston, MA 02110
Credit Cards Accepted: MC, VISA

◆ **The Weather Station**
What They Sell: Complete home weather stations plus other
weather gadgets. One of the better organized catalogs available
To Order, Call: (603) 526-6399
Shopping Hours: 10 AM to 4 PM; also have an answering
machine at other times
Or write to: The Weather Station, PO Box 1109, New London,
NH 03257
Credit Cards Accepted: MC, VISA, Discover

◆ **Wind & Weather**
What They Sell: Just about everything you want to know
about the weather—from home weather stations, instru-
ments, weathervanes, books, T-shirts, and more
To Order, Call: (800) 922-9463 or (707) 964-1284; Fax (707)
964-1278
Shopping Hours: 7 AM to 5 PM, Monday through Friday; 9 AM
to 5 PM, Saturday and Sunday
Or write to: Wind & Weather, PO Box 2320, Mendocino,
CA 95460
Credit Cards Accepted: MC, VISA, AMEX, Discover

174

Organizations

What better way to get into the weather than to join an organization of weather buffs. Here are two examples:

1. The American Weather Supplemental Observation Network (AWOSON). 401 Whitney Blvd., Belvidere, IL 61008-3772. Telephone: 8 AM-5 PM central, (815) 544-9811. After hours: (815) 544-3703.

One of the largest groups of weather buffs is AWOSON, and it's free to join! The group takes weather observations to supplement official records, although you don't have keep daily records to belong to the group. I like their monthly newsletter, *American Weather Observer*, which is stuffed with interesting weather stories. A one-year subscription is $16.95.

2. The Texas Severe Storms Association (TESSA). PO Box 122020, Arlington, TX 76012.

Despite its Texas roots, TESSA is a national organization "founded to bring both amateur and severe weather scientists together in an attempt to better understand severe storms." Annual membership is $14 for U.S. residents and $19 for international members. You get their quarterly newsletter, *The Weather Bulletin*, and invites to local meetings.

As a side note, the chairman of TESSA is none other than Martin Lisius, a well-known producer of documentaries on severe storms and tornadoes. I saw one of Martin's productions, "Surviving Tornado Alley," and was impressed. You can order this 21-minute tape, which highlights the hazards of tornadoes and lists the precautions necessary to survive them, for $20 (includes shipping) from the Plano Television Network, PO Box 860358, Plano, TX 75086-0358.

Do you know of an organization for weather buffs that's not listed above? Call me at (303) 442-8792 or write to the address at the front of this book, and I'll list it in the next edition of this publication.

About the Author

by Denise Fields

HAVE YOU EVER BEEN FORCED to sit through an entire hour of The Weather Channel? Or listen to minute by minute updates of outside temperature changes? How about waking up at 2AM to find your spouse in the dining room, watching a lightning storm in his underwear?

Well, I've experienced all of this and more. As the wife of a weather junkie, I admit to being a co-dependent as well. From the very day we met during the "Bronco Blizzard" of 1984, I have aided and abetted Alan's voracious appetite for all things weather. Sure, I gave him some of the gadgets he writes about in this book (including the car thermometer and the Casio weather watch), but I didn't know he'd write a whole book about this stuff.

Okay, maybe I should have known. Long before he met me while we were going to school at the University of Colorado, Alan was addicted to the weather. Although he was born in Chicago, he grew up in Dallas, Texas and was fascinated by the weather at a very early age. According to Alan's second-grade teacher, this obsession led him to issuing his own forecasts to classmates. He even "toured" other elementary schools to give weather talks.

While in college, Alan studied business and journalism, but I always found him drifting by the atmospheric sciences building. He finally broke down and took an introductory course in meteorology, which I suppose makes him more qualified to forecast the weather than most television weathercasters out there.

After graduating from CU in 1987, Alan went on to earn a Masters in Business Administration at the University of Texas at Austin. Now, at this point, you may wonder what a weather junkie is doing with all these business degrees. I'm not sure, but I think it had something to do with finding gainful employment—apparently, Alan planned to work at a real job during the day and dabble in the weather on the weekends.

Well, this plan went out the window in 1989. As two freshly minted college graduates, we found a job market that could be charitably described as abysmal. At this point, we turned to Plan B: why not become wildly successful book authors? Why not? We had the time.

Actually, this whole book thing was somewhat an accident. Our first book was a local guide to planning a wedding in Austin, Texas—something we just so happened to be doing while mailing 200 resumes a day. I think we sold a whopping 800 copies of that first book and were thrilled . . . at least we weren't digging ditches.

Oddly enough, this wedding book business took off dramatically in 1990. At that point, we wrote our first best-seller, a book called *Bridal Bargains*. This book was featured on the Oprah Winfrey Show in June 1991 and went on to sell over 150,000 copies. Suddenly, we were the country's foremost bridal experts—the weather junkie and I.

I should note that Alan's passion for the weather just got worse during the time we were writing about bridal gowns and wedding cakes. As we travelled the country to promote the book, we'd stop by local TV stations and do talk-shows with names like "Good Morning Boise." After each interview, Alan would disappear, only later to be found in the weather center, visiting with the TV weathercasters. He was like a kid in a toy store who never wanted to leave.

The next two books we wrote also had nothing to do with the weather—*Your New House* (about buying or building a new home) and *Baby Bargains* (about what you think it is about). Finally, we reached the point in our writing careers where Alan could write the book he's always dreamed of—the one you're holding in your hands right now.

So, what's next? Needless to say, Alan's next big project will be trying to interest our toddler in the weather. He started at an early age in Ben's life by meticulously recording the weather on the day Ben was born. Hopefully this means I won't be forced out into blizzards to check snow depths. Ben can take over. Maybe this parenting thing isn't half bad after all.

Index

(page entries in bold type indicate a primary discussion of topic)

A

A. C. Nielsen (TV ratings),40, 118

AC power, 110

Accu-Access for Windows (software), 141-142

Accu-Weather, 55, 59, 130, 138, 141, 144, 148

Accu-Weather Forecaster (software), 142

Accuracy of weather stations, 69, 70, 74, 82, 101

Affordability of weather stations, 65, 66

After Five Jazz, 47

Alarms, and weather stations, 71

Alarms for severe weather, 99, 101, 104, 153, 155, 160, 161

Alaska, 125

Albany, Georgia, 61

Albert the Alley Cat, 34-35

Alberta Clipper, 13

Alden Electronics, 125, 138, 140-141

Alert models, Maxon, 105, Midland, 105, Radio Shack, 105, WeatherOne, 106

Allen (hurricane), 157

Allrad (software for NEXRAD), 141

America On-Line, 140
  and downloading weather data, 130-131, 133, 144

American Meteorological Society (AMS), 10, 35

American Weather

Supplemental Observation Network (AWOSON), 175

Ames (Iowa), 58

Andrew (hurricane), 29-30, 139, 163

Anemometer, 69

Annen, Will, 53

Antioch (California), 21

Appalachians, 159

Arctic air, 1, 158, 160

Atlanta, 61

Atlantic Ocean, 20, 158, 159, 162
  air over, 1
  weather over, 123

Australia, 139

Author, telephone number of the, 97

Automated Weather Source, 71, **94-97**

Automatic Weather Stations (MesoTech), 89-90

Auxiliary temperature probe, 83

Aviation weather, downloading, 142, 143, 144

B

Bangor, 54

Barometers, 68

Basement, shelter, 156

Basic models of weather radios, Radio Shack, 106, Sony, 107

Beach forecast, 123

Beaty, David, 146

Beaverton (Oregon), 25

Bermuda High, 163

Beyond Police Call, 110
Blizzard of March 1993, 41
Blizzards, 19, 23, **158-161**
Blizzards, ground, 13
Blue screen, 44
Boating forecast, 123
Bob (hurricane), 23, 40-41
Boise, 157
Bono, Mike, 54
Books
    Beyond Police Call, 110
    Consumer Reports, 105
    Police Call, 110
    Television Weathercasting
      (Henson), 9
    Webster's New Geographical
    Directory, 44
    See also under Sources of
      information
Boston, 22-23, 124
Boulder, 74, **128**, 159
Brass instruments (Maximum),
  79-80
Brewick, Craig, 42
Broadcasts, National Weather
  Service, **100-104**
Brown, Jill, 54
Brown, Vivian, 54
Brownsville (Texas), 16
Brunotte, Gary, 47
Building regulations, and
  hurricanes, 163

C
California, 146, 158
Cameras, live, in major cities
  (The Weather Channel), 47
Canada, 13, 120, 124
  forecasts for, 132
  graphics shareware for, 133
Canadian air, 157, 159
Cannon, Declan, 55
Cantore, Jim, 42, 55
Capricorn weather stations
  (Hinds Instruments), 91-92
Career tracks, 53-61
Caribbean area, forecasts for,
  132
Caribbean Sea, weather over,
  123
Cascades, 159

Casem, Casey, 100
Catalogs
  American Weather
    Enterprises, 83, 112, 171
  Bass Pro Shops Marine
    Catalog, 171
  books, 171, 173, 174
  Casio Outdoorsman watch,
    173
  clothing, 174. *See also*
    The Weather Channel
  Edmund Scientific, 111,
    171
  experiments, 171
  Hammacher Schlemmer,
    106, 169, 171
  Heartland America, 75, 172
  instruments, 171-174
  Into the Wind, 172
  KLOCKIT, 172
  Museum Collections, 167
  Nature Company, 170, 172
  Northwest Discount
    Weather Catalog, 80, 92,
    173
  Radio Shack, 70, 105-107,
    109-110
  radios, 173. *See also*
    Radio Shack
  rain gauges, 173
  Robert White Weather
    Instruments, 174
  Safety Zone, 106, 169, 173
  Scanners, 172. *See also*
    Radio Shack
  Sharper Image, 169, 173
  Sporty's Pilot Shop, 173
  Sporty's Preferred Living,
    105, 173
  thermometers, 171-174
  videos, 173
  weather catalogs, 71-72,
    171-174
  weather gadgets, 171, 173
  Weather Station Catalog,
    80, 89, 174
  weather stations, 172-174
  Weatherama Weather
    Instruments, 174
  weathervanes, 174
  Wind and Weather, 105,

106, 169, 170, 174
wind meter, 172
Wireless catalog, 168
Categories of weather stations,
66-67
Cellular phones, 108, 109-110
and weather data, 148
Characteristics of good
weathercasters, 7, 43
Chattanooga, 21
Chesapeake Bay, 20
Cheyenne, 158
Chicago, 34-35, 157
Chroma key, 44
Claremont (New Hampshire),
55
Clothing, merchandising of, 48
Clowns as weathercasters, 9
CNN, 140
Cold front, 152
Coleman, John, 17, 35, 40
Colorado, 155, 158, 159-160
ColorGraphics, 28
Columbia (Missouri), 61
Columbus (Georgia), 21
Columbus Day storm
(Portland, Oregon), 23-24
Comfort index, 82
Commercial uses of weather
stations, 66, 67
CompuServe
and downloading weather
data, 131-132, 144
and shareware, 133
Computer connections, and
weather stations, 71
Computers
Accu-Access for Windows
(software), 141-142
Accu-Weather Forecaster
(software), 142
Allrad (software for
NEXRAD),141
America On-Line, 130-131,
133, 140, 144
Canada, forecasts for, 132
Canada, graphics
shareware for, 133
Caribbean area,
forecasts for, 132
ColorGraphics, 28

CompuServe, and
downloading weather
data, 131-132, 144
CompuServe, and
shareware, 133
connect charges 138, 140,
142, 143, 144, 148
Contel DUAT (Direct User
Access Terminal), and
downloading weather
data, 132
cost of downloading, 138,
140, 142, 143, 144, 148
Data Transformation
Company, and
downloading weather
data, 132
and Davis weather
stations, 75-76
Delphi, and access to the
Internet, 133
downloading graphics and
forecasts, 130-132, 157
Europe, graphics shareware
for, 133
Flight Data, 144
and Fourth Dimension
system, 83-86
gateways to the Internet,
133
GEnie, and downloading
weather data, 132
graphics, 10, 141, 143
Grove Enterprises (weather
fax), 146
Haley, Keith, 129
Haywood, Jim, 133
hurricane software, 163
Hurrtrack (software),
162-163
IBM. *See* IBM-PC
Illinois, University of,
weather bulletin board
on the Internet, 132
Internet, 132-133
K&H Enterprises, 129
lightning analysis software,
112
"Liveline1," 28
Macintosh, 71, 76-77, 95,
127, 129, 137, 139

McCallie Corporation, and
   lightning analysis
   software, 112
models, problems with, 19
modem, 95, 130, 132, 133,
   143
National Hurricane Center,
   forecast data from, 163
OFS Weather FAX, 146
Ohio State University,
   weather bulletin board
   on the Internet, 132
on-line services and The
   Weather Channel, 49
PC HF Fax, 146
PC Weatherman, 144
Power Macintosh, 137
Prodigy, and downloading
   weather data, 132
and radar pictures, 101
RS232, 79, 82, 88, 89, 90,
   91, 93
RS422, 90
Satellite Data Systems
   (weather fax), 146
and screen phones, 49
"Skyscape computer,"
   27-28
statistical forecast track
   (hurricane data
   software), 163
Universal Radio
   (weather fax), 146
Vasquez, Tim, 134
voice synthesis, 50
"Weather Channel
   Forum," 49
weather almanac
   (software), 129
weather bulletin boards
   on the Internet, 132
Weather Channel,
   graphics on, 45
Weather Channel,
   Three-dimensional
   maps on, 45
Weather Dimensions,
   software by, 85
Weather Network, 143-144
Weather Pro 5.5 (software),
   129

and Weather Report
   weather station, 82
Weather Star, 39, 50, 121
and weather stations, 71
Weather Watch (software),
   144
Weatherbrief (software),
   143
WeatherGraphix, 134
Weatherlink (software),
   75-77, 80
WeatherStat (software),
   129
WeatherView (shareware),
   133, 134
Concord (California), 21
Condella, Vince, 11-12
Connect charges, 138, 140,
   142, 143, 144, 148
Construction techniques, and
   hurricanes, 163
Consumer Reports, 105
Contel DUAT (Direct User
   Access Terminal), and down-
   loading weather data, 132
Continental Divide, 160
Costs
      of downloading weather
         data, 138, 140, 142, 143,
         144, 148
      of weather stations, for
         educational use, 94-97
      of weather stations, for
         home use, 72-89
      of weather stations,
         for industrial and
         professional use, 89-94

D
Dahl, Dave, 13-15
Dallas, 7, 15-16, 156, 158
Data Transformation
   Company, and downloading
   weather data, 132
Davis rain collector, 71
Davis weather stations, **72-77**,
   127
Death toll, 18, 110, 155, 168
Debardelaben, Bob, 18
Debugging, initial, at The
   Weather Channel, 39-40

Degree days heating and
  cooling, 86
Delphi, and access to the
  Internet, 133
Denver, 13, 27, **28-29**, 158
  "Denver Bronco Effect,"
  160
Des Moines, 58
Desert Southwest, 157
Dew point, 68, 153
Digital weather stations, 65
Dixie Dart, 13
Doppler radar, 111, 117, 138.
  148
Downbursts, 152
Downdrafts, 152
Downloading graphics and
  forecasts, 130-132, 157
Dry Line, 152
Dry thunderstorms, 152
Dust devils, 155

E
East Coast, 158, 159, 163
Eck, Dale, 55
Education
  donations for, 96
  programming, 123
  uses of weather stations, 67
  "weather talks," 7, 28
Edwards, Brad, 56
EKO, 48
Emergency management, and
  weather stations, 67
Emergency weather
  broadcasting, Hurricane
  Andrew, 30
Emily (hurricane), 146
Environmental Protection
  Agency, 17
Ethnic diversity, 42
Europe, 139
  forecast for, 123
  graphics shareware for, 133
Evansville (Indiana), 59
"Exit," on weather information
  superhighway, 117

F
Fairbanks, 167
Fascinating Electronics, Inc.,
128
Faults of weather stations,
  73, 74, 78-79, 80, 86-88, 89,
  91-92, 93-94
Features of weather stations.
  See Functions
Finfrock, David, 15-16
Fire departments, 108
Fishel, Greg, 17-18
Flagstaff, 32, 61
Flight Data, 144
Flooding, 12
Florida, 111
Fog, 21
Forecast Center
  (The Weather Channel), 41
Forecasting
  downloading, 142
  method of, 22-23
  perils of, 16, 19
Forecasts, discussions of, via
  Weather Network, 144
Forest fires, on satellite images,
  146
Fort Pierce (Florida), 56
Fort Worth, 15-16, 156
Fourth Dimension History
  Logging Weather Station,
  83-86
  and Fourth Dimension
    system (software), 83-86
Freezing rain, 24, 159, 160
Frequencies
  National Weather Service
    (NMS), 99, 100
  of various entities, 110
Front Range, 159, 160
Frost forecast, 65
Fujita Wind Damage Scale, 154
Functions, **67-71**
  Automatic Weather
    Stations, 89-90
  Capricorn weather
    stations, 91-92
  Fourth Dimension History
    Logging Weather
    Station, 83-86
  Nimbus instruments, 92-94
  Observer, 128
  Perception II, 75
  Rainwise, 86-88

Ultimeter II, 88-89
Weather Monitor II, 72-73, 76
Weather Report, 81
Weather Wizard III, 75, 77, 127
WeatherMAX, 77-81
Future of The Weather Channel, 48-50, 147-148

G

Galumbeck, Alan, 39-40, 48, 49, 50
Gardeners, and the weather, 65-66, 88
Gateways to the Internet, 133
Gender of forecasters, 42
GEnie, and downloading weather data, 132
Gibson, Charlie, 26
Gifts for weather junkies, 167-170
  Alaska Weather Calendar, 167
  Car thermometer and clock, 168-169
  Casio Outdoorsman watch, 169
  Clothing, 168
  Galileo thermometer, 169-170
  Kansas Storms: Destruction, Tragedy, & Recovery 1991, 168
  Nature Company gift certificate, 170
  Pacesetter Six Weather Station, 169
  Umbrella, 167
  Weatherwise magazine, 168
Gloria (hurricane), 23
"Good Morning America," 17, 26, 27, 35, 40
Graphics, 50
  early, 10
  modern, 141, 143
Great Plains, 111, 157, 159
Green, Jim, 167
Green Bay, 53
Green sky, hail and 37

Greenville, North Carolina, 60
Griffin, Rick, 56
Grove Enterprises (weather fax), 146
Gulf of Mexico, 120, 157, 159, 160, 162
  air over, 1, 157
  weather over, 123
Gustnadoes, 154

H

Hail, 37, 151, 152
Haley, Keith, 129
Ham radio, 108
Hartford, 54
Hastings (Nebraska), 59
Hawaii, 125
Haywood, Jim, 133
Heat waves, 156-158
Henson, Robert, 9
Highway travel, weather affecting, 123, 124, 148, 161
Hill, Doug, 19-20
Hinds Instruments, 91-92
Hoitsma, Chris, 47
Home front, and weather stations, 66, 158
Homestead (Florida), 139
Hope, John, 42-43, 56
Huff, Janice, 20-21
Humidity, 68, 78-79, 153
Hurricanes, 23, 29-30, 69, 140, 142, 146, **161-163**
  Allen, 157
  Andrew, 29-30, 139, 163
  Bob, 23, 40-41
  building regulations, 163
  construction techniques, 163
  Emily, 146
  Gloria, 23
  Miami Herald article on Hurricane Andrew, 163
  and weathercasters, 29-30, 60
  and West African coast, 161
Hurrtrack (software), 162-163
Husband and wife weather team, 56
Hyperscan, 109

I

IBM PC-compatible, 71, 76-77,
79, 81, 82, 83, 85, 90, 93, 95,
112, 127, 128, 129, 143, 144,
147, 163
with fast processor and
high resolution monitor,
137, 141, 143
386 or greater, 139, 141
Ice storms, 158, 159, 160
Illinois, University of, weather
bulletin board on the
Internet, 132
Improvements to The Weather
Channel, suggestions for,
45-47, 120
Indianapolis, 124
Information boxes, and
weather stations, 77, 80-81,
83, 88, 90, 92, 94, 97
International weather, 119,
122-123
Internet, 132-133

J

Jet stream, 18, 157
Johnson, Rich, 57
Jones, Jeanetta, 57
Joplin (Missouri), 61

K

K&H Enterprises, 129
Kansas City, 36-37, 157
Keneely, Bill, 57
Kinds of weather station
measurements, **67-71**

L

Lake effect, 12, 69, 159
Lake Michigan, 12
Landmark Communications,
40
Landmark Video Networks and
Enterprises, 48
Landspouts, 154, 155
Lemke, Cheryl, 58
Leonard, Harvey, 22-23
Light sensor, 71
Lightning, 152
and casualties and damage,
110-111
detectors, 110-112, 153
and downloading data,
142, 144
software for counting
strikes, 112
Lightning rod, 70
Linquist, Terry, 142
Lisius, Martin, 175
Little, Jim, 23-25
"Liveline1," 28
Local climate
Boston, 22
Boulder, 74, 128
Denver, 28-29,
Kansas City, 36-37
Milwaukee, 12
North Texas, 16
Phoenix, 30-31
Portland, 24
Raleigh/Durham, 18
San Francisco, 21
St. Louis, 26-27
Washington, D.C.,19-20
Local forecasts, 39, 118-124
downloading, 138
talking, 49-50
See National Weather
Service, 100-101.
See "Talking yellow pages,"
121, 124
See Weather Channel, The,
39, 49-50, 65, **122-124**.
Logs. *See* Records
Long-range forecasts, 45-46,
119, 120, **121-124**
by fax, 124
Lunden, Joan, 26
Lydon, Mike, 90
Lyons, Dr. Walt, 14

M

Macintosh, 71, 76-77, 95, 127,
129, 137, 139
Macon, 55, 57
Maestro, 77
Maine, 159
Major cities, outlook for, 123
Mancuso, Mark, 58
Manitoba Mauler, 13
"Map blockage," 42
Maps in motion, 119, 120

Marine transmissions, 108
Marine weather, downloading, 142
Market-driven forecasts, 42
Maximum's WeatherMAX. *See* WeatherMAX
Maxon, 100, 105
McCallie Corporation, and lightning analysis software, 112
Measurements, weather stations, **67-71**
Media, and frequencies used by, 108-109
Media market, 159
Memphis, 60
Merchandising, and The Weather Channel, 48
Mesonet (weather forecasting company), 14
Mesoscale convective complexes, 37
MesoTech, 89-90
Miami, 29-30
Miami Herald article on Hurricane Andrew, 163
Microclimate, 21. *See also* Local climate.
Military Shortwave Radio Weather Facsimile network, 145-146
Military weathercasters, 9, 24, 36
Milwaukee, 11
Minneapolis, 13-14
Mississippi River, 13
Models for forecasting, problems with, 19
Modem, 95, 130, 132, 133, 143
Monsoon, 30-31
Montana, 159
Moore, Thomas, 58
Morrow, Jeff, 59
Mountains, effects of, 19, 24, 28-29
Murray, Dave, 25-27
Music (The Weather Channel), 47-48

N
National Hurricane Center, forecast data from, 163
National Weather Service (NMS), 14, 17, 110-111, 139
broadcasts and warnings by, 100, 153, 155, 160
data via computer from, 69, 143, 144
forecasts by, 11, 45, 65, 103, 121, 161
frequencies, 99, 104, 148
future of, 148
observations of, 66
suggestions for, 102
telephone number for, 124
trainee for, 21
weather spotters, 108
Nebraska, 160
Nelson, Mike, 27-29
Nenana, 167
New England, 156, 159
New Mexico, 111, 159
New York City, 2, 124, 159
NEXRAD (Next Generation Weather Radar), 12, 46, 138, 140, 141, 142, 144, 155, 161
Nielsen, A. C., Company (TV ratings), 118
Nimbus instruments, 92-94
Nor'Easter, 158
Norcross, Bryan, 29-30
North Carolina, 146
North Dakota, 111
North Pole, 139

O
Observer, 128
OFS Weather FAX, 146
Ohio State University, weather bulletin board on the Internet, 132
Oklahoma, 155, 156
Oklahoma City, 9
Oklahoma Panhandle, 13
Olympic Mountains, 33
On-line services and The Weather Channel, 49
Oracle, 87-88
Oregon, 159
Oregon Scientific, 169
Oswego (New York), 58
Outdoor activities, weather

affecting, 123
Ozone, forecast of , 21

P

Pacesetter Six Weather Station, 75, 169
Pacific Northwest, 25, 33
Pacific Ocean air, 1
Panhandle Hooker, 13
Parents' influence, 14
PC HF Fax, 146
PC Weatherman, 144
Perception II, 75
Phillips, Ed, 30-32
Phoenix, 30-31, 157
Pilots, 125. *See also* Aviation weather
Plano (Texas), 107, 175
Police, 107, 109, 155
Police Call, 110
Pollen, report of, 21
Pollution, report of, 21
Pool, Steve, 32-33
Popcorn thunderstorms, 151-153
Portland (Oregon), 23-25
Portsmouth, Virginia, 61
Power Macintosh, 137
Prodigy, and downloading weather data, 132
Pronunciation of place names, 44
Puerto Rico, 125
Puget Sound Convergence Zone, 33
Puppets as weathercasters, 9, 35

R

Race of forecasters, 42
Radar, 50, 103. 119, 121, 122, 155
    images, 101, 139, 141, 143, 144
    and wind direction, 12, 46, 138
Radios
    AC power, 110
    alarms for severe weather, 99, 101, 104, 153, 155, 160, 161

alert models, Maxon, 105, Midland, 105, Radio Shack, 105, WeatherOne, 106
basic models, Radio Shack, 106, Sony, 107
battery option, 104, 110
broadcasts, National Weather Service, **100-104**
cellular phones, 108, 109-110
fire departments, 108
frequencies, National Weather Service (NMS), 99, 100
frequencies, of various entities, 110
ham radio, 108
hyperscan, 109
lightning detectors, 110-112, 153
marine transmissions, 108
Maxon, 100, 105
McCallie Corporation, 112
media, and frequencies used by, 108-109
police, 107, 109, 155
Radio Shack, 104, 145
scanners, **107-110**, 155, 161
shortwave, 137, 145-146
State Patrol, 108, 161
Storm Alert (lightning detector), 111
Stormwise Lightning Alert (McCallie), 112
tornadoes, 101, 104, 155
wake-up feature, 106
warnings, 102-103, 155, 160
watches, 101-103, 155, 160
weather spotters, 108
Rain collectors, 70
Rain gauges, 70
Rainwise sensors, 71, 83
Rainwise. *See* Oracle, WeatherStation, WeatherVideo
Raleigh/Durham, 17-18
Ramsay, Ray, 32-33

Records and logs, 71, 73, 76, 80, 84-86, 123, 127

Refraction, 70

Requirements for weather station evaluations, 66-67

Road conditions, 124, 148, 161

Rocky Mountains, 111, 158, 159

Rottman, Leon "Stormy," 28

RS232, 79, 82, 88, 89, 90, 91, 93

RS422, 90

Ryan, Bob, 19, 26

S

Sacramento, 24

Saeland, Jodi, 59

Saffir-Simpson Scale, 162

Salaries (The Weather Channel), 43

Salt Lake City, 16, 24, 124, 157

San Francisco, 20-21

Satellite Data Systems (weather fax), 146

Satellite images, 10, 22, 117, 119, 138, 139, 140, 141, 142, 143, 144, 145

Sawyer, Diane, 10

Scanners, **107-110**, 155, 161

Scanning display, and weather stations, 73

Schedule (The Weather Channel), 119, **122-123**

Schools, and weather stations, 67

Schwartz, Dave, 59

Screen phones, 49

Sea salt, and weather stations, 69

Seattle, 32, **33**, 124

Seidel, Mike, 60

Sensors
with no moving parts, 89
placement of, 67-68, 70

Shadowfax, 47

Shahin & Sepehr, 48

Shelter during tornadoes, 154, 155

Shortwave, 137, 145-146

Sierra Nevada, 158, 159

"Silver fog," 24

Single-cell thunderstorm, 151

Sioux City, Iowa, 61

Ski resorts, and the weather, 66, 124

Skilling, Tom, 34-36

"Skyscape computer," 27-28

Smith, Dennis, 60

Smith, Steve, 47

Smith, Terri, 61

Snow
amount projections, 160
conditions, 124
heavy, 158
lake effect, 159

Software, 168.
See also individual listings

Sources of information
Accu-Weather, 141
Alden Electronics, 125, 140
America On-Line, 131
American Weather Supplemental Observation Network (AWOSON), 175
American Weather Observer, 175
Automated Weather Source, 97
books, 9, 168, 171, 173
CompuServe 130
Contel DUAT (Direct User Access Terminal), 132
Data Transformation Company, 132
Davis Instruments, 77
Fascinating Electronics, Inc., 128
GEnie, 132
Grove Enterprises (weather fax), 146
Hearth Publishing, 168
Heldref Publications, 168
Hinds Instruments, 92
K&H Enterprises, 129
Maximum, 80
McCallie Corporation, 112
McFarland & Company Publishers, 9
OFS Weather FAX, 146
Ordering information.
See also Catalogs.

Plano Television Network, 175

PC Weather Products, 162-163

Prodigy, 132

Rainwise, 88

Satellite Data Systems (weather fax), 146

Say Soft, 144

Sensor Instruments, 94

Software Systems Consulting 145-146

Telonics, 146

Texas Severe Storms Association (TESSA), 134, 175

Texas Weather Instruments, 83

Universal Radio (weather fax), 146

Vanguard Electronic Labs, 146-147

Weather Bank, 143

Weather Bulletin, The, 175

Weather Channel, The, 48

Weather Dimensions, 86, 127

Weather Network, 143

Weather Services International Corporation, 139

Weather View, 133

Weatherwise, 146, 168

Williwaw Publishing Company, 167

ZFX, 125

South, The, 158

South Burlington (Vermont), 55

Southwesterly winds, and heat build-up, 157, 158

Spartanburg (South Carolina), 57

Spencer, Lisa, 60

Sports-related items, merchandising of, 48

Squall line, 151, 152

St. Louis, 21, 25, **26-27**

Stanier, Marny, 44, 61

State Patrol, 108, 161

Statistical forecast track (hurricane data software), 163

"Storm of the Century" (1993), 160

Storm Alert (lightning detector), 111

Storm track, 157

Stormwise Lightning Alert (McCallie), 112

Story aspect of weather, 43

Suggestions for National Weather Service, 102

Suggestions for The Weather Channel, 45-47, 120

Summer from Hell (1980), 156

Sunshine index, 71

Super-cell storms, 151, 152-153

Surf conditions, 124

Surge protector, 70, 111

T

T-Ris Remote Imaging System, 146

Taft, Harold, iii, 15-16

"Talking yellow pages," 117, 121, 124

Tampa, 57

Tanan River, 167

Telephone, weather forecast on, 121

Telephone numbers
   of the author, 97
   of The Weather Channel, 49, 124

Television stations
   KATU-TV (Portland, Oregon), 24
   KBTV-TV (Denver), 28
   KCAU-TV (Sioux City, Iowa), 61
   KCMO-TV (Kansas City, Missouri), 36
   KCTV-TV (Kansas City, Missouri), 36
   KGW-TV (Portland, Oregon), 23, 24
   KHAS-TV (Hastings, Nebraska), 59
   KMOX-TV (St. Louis), 28
   KNAZ-TV (Flagstaff), 61
   KNXV-TV (Phoenix), 30-31

KOIN-TV (Portland, Oregon), 23, 24
KOKH-UHF (Oklahoma City), 36
KOMO-TV (Seattle), 32
KOY-AM (Phoenix), 31
KPNX-TV (Phoenix), 31
KRON-TV (San Francisco), 20
KSAZ-TV (Phoenix), 31
KSD-TV (St. Louis), 25-26
KSTP-TV (Minneapolis), 13-14
KTVI-TV (St. Louis), 25
KUSA-TV (Denver), 27
KWTV-TV (Oklahoma City), 36
KXAS-TV (Dallas/Ft. Worth), 15
WALF-TV (Albany, Georgia), 61
WAVY-TV (Portsmouth, Virginia), 61
WBZ-TV (Boston), 25-26
WCAX-TV (South Burlington, Vermont), 55
WDAF-TV (Kansas City, Missouri), 36
WEVV-TV (Evansville, Indiana), 59
WFSB-TV (Hartford), 54
WGN-TV (Chicago), 34-36
WGXA-TV (Macon), 55
WHBQ-TV (Memphis), 60
WHDH-TV (Boston), 22
WHDQ (Claremont, New Hampshire), 55
WITI-TV (Milwaukee), 11, 34
WKIV-TV, 56
WKKD-AM (Aurora, Illinois), 34
WKOW-TV (Madison), 34
WLBZ-TV (Bangor), 54
WLUK-TV (Green Bay), 53
WLXT-TV (Aurora, Illinois), 34
WMAZ-TV (Macon), 57
WPLG-TV (Miami), 29
WRAL-TV (Raleigh/ Durham), 17
WRC (Washington, D.C.), 19
WSPA-TV (Spartanburg, South Carolina), 57
WTHI-TV (Terre Haute, Indiana), 58
WTKA-TV (Topeka), 56
WTSO-AM (Madison), 34
WTVJ-TV (Miami), 29, 30
WTVT-TV (Tampa), 57
WTVX (Ft. Pierce, Florida), 56
WUSA-TV (Washington, D.C.), 19
WXIA-TV (Atlanta), 61
WXY-TV (Oklahoma City), 9
WXYZ (Detroit), 19
WYFF-TV (Greenville, North Carolina), 60
Television Weathercasting (Henson), 9
Temperature, 67
Ten-day forecast, 140
Terre Haute (Indiana), 58
Texas, 16, 111, 152, 156, 157, 175. *See also* individual cities.
Texas Panhandle, 13
Texas Severe Storms Association (TESSA), 134, 175
Texas Weather Instruments, **81-83**
Thermal Field Variation, 89
Thermal low, 31
Third Force, 47
Thompson, Mike, 36-37
Thunderstorms, 20, **151-153**, 161
    "disappearing," 26-27
    fear of, 20
    and lake effect, 12
    and mountain effect, 29
    and Phoenix, 31-32
    prediction of, 68
Tide clock, 82
Tide conditions, 124
Topeka, 56
Tornado alley, 36, 107, 175
Tornado warnings, Plano, Texas, 107

Tornado watches, 101,
Tornadoes, 104, 151, 152,
  153-156
    fear of, 17
    and Midwest, 104
    and NEXRAD, 46
    and Plains, 29, 104, 168
    and Raleigh/Durham,
      13, 18
    and Twin Cities, 13
Trend of barometric pressure,
  68, 74, 75
Tropical cyclone, 162
"Tropical Update," 56, 122, 162
Tropics, 44
Trump, Donald, 146
TV audition, 11, 15-16, 25
Twain, Mark, iii
Twin Falls (Idaho), 24

U
Ultimeter II, 88-89
Units of measurement, 69, 71,
  81
Universal Radio (weather fax),
  146
Universities and colleges with
  meteorology programs
    Chicago, 56
    Florida State, 14, 21, 29,
      57, 61
    Georgia, 57
    Iowa State, 56, 58
    Jackson State, 54
    Lyndon State College,
      55, 56
    Maryland, 19, 58
    Maryland State, 19
    Memphis State, 60
    New Hampshire, 57
    New York, City College of,
      22
    New York University, 22,
      54
    North Carolina State, 17
    Oklahoma, 60
    Penn State, 55, 58, 59
    Purdue, 11
    St. Louis, 31
    Temple, 59
    Texas A&M, 15
    Texas in Austin, 24
    Washington, 33
    Wisconsin, 11, 34, 53, 59
    Wyoming, 25
Update hotline, 149
Updrafts, 152, 153
Upslope snow, 159-160
Urban heat-island effect, 27
U.S. Virgin Islands, 125
Utility companies, and the
  weather, 66

V
Van Roy, Jack, 7-8
Vasquez, Tim, 134
Video "on demand," 48-49
Videos, merchandising of, 48,
  168, 175
Virga, 152
Voice synthesis, and The
  Weather Channel, 50
Volunteer fire departments,
  and weather stations, 80

W
Wake-up feature, 106
Warnings, 102-103, 121, 155,
  160
Warranty on weather station,
  92
Washingtion (state), 159
Washington, D.C., 19-20
Watches, 101-103, 155, 160
Weather almanac, 31-32
Weather almanac (software),
  129
Weather briefings, 43
Weather bulletin boards on the
  Internet, 132
Weather by phone, 121
Weather Channel, The,
  17, 35, 39-50, 117, 118-121,
  137, 140, 145, 147-148, 153,
  157, 160
    A. C. Nielsen (TV ratings),
      40, 118
    After Five Jazz, 47
    Annen, Will, 53
    Atlantic Ocean, weather
      over, 123
    Blizzard of March 1993, 41

blue screen, 44
Bono, Mike, 54
Brewick, Craig, 42
Brown, Jill, 54
Brown, Vivian, 54
Brunotte, Gary, 47
cameras, live, in major
    cities, 47
Cannon, Declan, 55
Cantore, Jim, 42
Caribbean Sea, weather
    over, 123
chroma key, 44
Coleman, John, 40
debugging, 39-40
Eck, Dale, 55
educational programming,
    123
Edwards, Brad, 56
EKO, 48
ethnic diversity, 42
Forecast Center, 41
future of, 48-50, 147-148
Galumbeck, Alan, 39-40,
    48, 49, 50
gender of forecasters, 42
graphics on, 45
Griffin, Rick, 56
Gulf of Mexico, weather
    over, 123
highway travel, weather
    affecting, 123
Hoitsma, Chris, 47
Hope, John, 42-43, 56
improvements, suggestions
    for, **45-47**, 120
international weather, 119,
    122-123
Johnson, Rich, 57
Jones, Jeanetta, 57
Keneely, Bill, 57
Landmark
    Communications, 40
Landmark Video Networks
    and Enterprises, 48
Lemke, Cheryl, 58
local forecasts, 39, 49-50,
    118-124
local forecasts, talking,
    49-50
long-range forecasts,

    45-46, 119, 120, **121-124**
major cities, outlook for,
    123
Mancuso, Mark, 58
"map blockage," 42
maps in motion, 119, 120
market-driven forecasts, 42
merchandising, 48
Moore, Thomas, 58
Morrow, Jeff, 59
music, 47-48
NEXRAD, 46
Nielsen, A. C., Company
    (TV ratings), 118
on-line services and, 49
outdoor activities, weather
    affecting, 123
pronunciation of place
    names, 44
race of forecasters, 42
records of heat and/or
    humidity, 123
Saeland, Jodi, 59
salaries, 43
schedule, 119, **122-123**
Schwartz, Dave, 59
screen phones, 49
Seidel, Mike, 60
Shadowfax, 47
Shahin & Sepehr, 48
Smith, Dennis, 60
Smith, Steve, 47
Smith, Terri, 61
snow accumulation, 160
Spencer, Lisa, 60
Stanier, Marny, 44, 61
story aspect of weather, 43
telephone numbers,
    49, 124
Third Force, 47
three-dimensional maps
    on, 45
"Tropical Update," 56, 122
    162
tropics, 44
video "on demand," 48-49
warnings, 121
weather briefings, 43
"Weather Channel
    Forum," 49
"Weather on demand,"

148-149
weather principles, 123
Westerlage, Keith, 44, 61
Weather Dimensions,
    software by, 85
Weather experiment kits, 14
Weather gadgets, 14
Weather girls, 9-10
Weather instruments,
    merchandising of, 48
Weather Monitor II, **72-74**, 97,
    127, 172
Weather Network, 143-144
    and forecasts discussions,
        144
"Weather on demand,"
    148-149
Weather phenomena
    importance of, 2-3
    reasons for, 1
    See individual listings
Weather principles, 123
Weather Pro 5.5 (software), 129
Weather Report (weather
    station), 81-83
Weather Services International
    Corporation (WSI), 138, 139,
    144, 148
Weather spotters, 108
Weather Star, 39, 50, 121
Weather stations
    accuracy, 69, 70, 74, 82,
        101
    affordability, 65, 66
    alarms, 71
    anemometer, 69
    Automated Weather
        Source, 71, **94-97**
    Automatic Weather
        Stations (MesoTech),
        89-90
    auxiliary temperature
        probe, 83
    barometers, 68
    brass instruments
        (Maximum), 79-80
    Capricorn weather stations
        (Hinds Instruments),
        91-92
    categories, **66-67**
    comfort index, 82

commercial uses, 66, 67
computer connections, 71
costs, for home use, 72-89
costs, for industrial and
    professional use, 89-94
costs, for educational use,
    94-97
Davis rain collector, 71
Davis weather stations,
    72-77, 127
degree days heating and
    cooling, 86
dew point, 68
digital, 65
educational uses, 67
emergency management,
    67
Fascinating Electronics,
    Inc., 128
faults, 73, 74, 78-79, 80,
    86-88, 89, 91-92, 93-94
features. *See* Functions
Fourth Dimension History
    Logging Weather
    Station, 83-86
Hinds Instruments, 91-92
    and the home front,
        66, 158
humidity, 68, 78-79
information box, 77, 80-
    81, 83, 88, 90, 92, 94, 97
kinds of measurements,
    67-71
light sensor, 71
lightning rod, 70
logs. *See* Records
Maestro, 77
Maximum's WeatherMAX.
    See WeatherMAX
measurements, **67-71**
MesoTech, 89-90
Nimbus instruments, 92-94
Observer, 128
Oracle, 87-88
Oregon Scientific, 169
Pacesetter Six Weather
    Station, 75, 169
Perception II, 75
    and personal computers,
        65, 71.
        See also Computers

rain collectors, 70
rain gauges, 70
Rainwise sensors, 71, 83
Rainwise. *See* Oracle,
   WeatherStation,
   WeatherVideo
records and logs, 71, 73,
   76, 80, 84-86, 127
refraction, 70
requirements, 66-67
RS232 79, 82, 88, 89, 90
RS422, 90
scanning display, 73
schools, 67
sea salt, 69
sensors, placement of,
   67-68, 70
sensors, and no moving
   parts, 89
sunshine index, 71
surge protector, 70
temperature, 67
Texas Weather
   Instruments, **81-83**
Thermal Field Variation,
   89
tide clock, 82
trend of barometric
   pressure, 68, 74, 75
Ultimeter II, 88-89
units of measurement,
   69, 71, 81
warranty, 92
Weather Monitor II, **72-74**,
   97, 127, 172
Weather Report, **81-83**
Weather Wizard III, 75
Weatherlink software,
   75-77, 80, 127
WeatherMaster software,
   91-92
WeatherMAX, 73, **77-81**
WeatherStation, 86-88
WeatherVideo, 87-88
wind chill, 69
Weather Watch (software), 144
Weather Will, 53
Weather Wizard III, 75
Weather-by-fax, 125, 141
Weatherbrief (software), 143
Weathercasters

Annen, Will, 53
Bono, Mike, 54
Brown, Jill, 54
Brown, Vivian, 54
Cannon, Declan, 55
Cantore, Jim, 42, 55
Coleman, John, 17, 35, 40
Condella, Vince, 11-12
Dahl, Dave, 13-15
Debardelaben, Bob, 18
Eck, Dale, 55
Edwards, Brad, 56
Finfrock, David, 15-16
Fishel, Greg, 17-18
Griffin, Rick, 56
Hill, Doug, 19-20
Hope, John, 56
Huff, Janice, 20-21
Johnson, Rich, 57
Jones, Jeanetta, 57
Keneely, Bill, 57
Lemke, Cheryl, 58
Leonard, Harvey, 22-23
Little, Jim, 23-25
Mancuso, Mark, 58
Moore, Thomas, 58
Morrow, Jeff, 59
Murray, Dave, 25-27
Nelson, Mike, 27-29
Norcross, Bryan, 29-30
Phillips, Ed, 30-32
Pool, Steve, 32-33
Ramsay, Ray, 32-33
Rottman, Leon "Stormy,"
   28
Ryan, Bob, 19, 26
Saeland, Jodi, 59
Schwartz, Dave, 59
Seidel, Mike, 60
Skilling, Tom, 34-36
Smith, Dennis, 60
Smith, Terri, 61
Spencer, Lisa, 60
Stanier, Marny, 44, 61
Thompson, Mike, 36-37
Van Roy, Jack, 7-8
Westerlage, Keith, 44, 61
Weathercasters' characteristics,
   7
WeatherGraphix, 134
Weatherlink software, 75-77,

80, 127
WeatherMaster software, 91-92
WeatherMAX, 73, **77-81**
WeatherStat (software), 129
WeatherStation, 86-88
WeatherVideo, 87-88
WeatherView (shareware),
    133, 134
*Webster's New Geographical
    Directory*, 44
Welsh, Mark, 144
West Coast, 159
West Texas, 152
Westerlage, Keith, 44, 61
Wichita Falls, 156

Wilmington, 18
Wind chill, 69
Wind Damage Scale, Fujita,
    154
Wind shear, 155
Windstorms, 23-24, 29, 69, 158
Wooly lamb, 9
Wyoming, 111

Y
Young forecasters, 34

Z
ZFX (weather-by-fax), 125, 141

"Far & Away Weddings"

Want to get married on a tropical beach?

How about in an 800-year old castle?

Or do you want to tie the knot back in a home-town that's across the country? How can you plan a wonderful wedding "long distance"? Find the answers in this new, paperback guide by the authors of "Bridal Bargains"! Inside this 225 page book, you'll learn:

- THE INSIDE SCOOP on over 25 exotic spots to tie the knot!
- SURPRISING STRATEGIES FOR FINDING WEDDING STORES and services "long distance."
- TRICKS FOR DEALING WITH "SURROGATE PLANNERS" (Mom and other well-meaning friends and relatives).
- WHAT YOU CAN DO IN YOUR "HOME CITY"—it's more than you think!
- IN-DEPTH "BATTLE PLANS" for those brief trips to the "wedding city"—suggested schedules for meeting with the most wedding merchants in the shortest period of time!
- CREATIVE IDEAS FOR COORDINATING far-flung bridesmaids and groomsmen!
- TIPS ON HIRING and working with wedding consultants and planners.
- INNOVATIVE WAYS TO USE TECHNOLOGY such as fax machines and video cameras to ease the planning process.
- And, of course, COST-CUTTING TIPS for holding down expenses!

Just

$8.95

(Plus $3 shipping)

Call toll-free to order!
1-800-888-0385

Mastercard, VISA, American Express and Discover Accepted!

"Baby Bargains"

Secrets to saving 20% to 50% on baby furniture, equipment, clothes, toys, maternity wear and much, much, more!

Hooray! A baby book that actually answers the big question about having a baby: How am I going to afford all this? With the average cost of a baby topping $5000 for just the first year alone, you need creative solutions and innovative ideas to navigate the consumer maze that confronts all parents-to-be. Baby Bargains is the answer! Inside, you'll discover:

- How to Save Up to 25% on brand-new, designer-brand baby bedding.
- Five Wastes of Money with baby clothes and which brands are the best.
- Seven tips to Saving Money on Cribs, plus in-depth reviews of the top crib makers.
- How to get a name-brand car seat at Wholesale!
- The Truth About Strollers—and which brands work best in the real world.
- Dozens of Safety Tips to affordably baby proof your home.
- The Top 10 Best Baby Gifts and five gift don'ts.
- Name Brand Reviews of toys, monitors, high chairs, diapers and more!
- Seven Creative Sources for maternity clothes—and a national outlet that offers 40% savings!

Money-Back Guarantee: If *"Baby Bargains"* doesn't save you at least $250, we will give you a complete refund. No kidding!

Just
$11.95
(Plus $3 shipping)

Call toll-free to order!
1-800-888-0385
MasterCard, VISA, American Express and Discover Accepted!

How to Reach the Author

Have a question about

Partly Sunny?

*Want to make
a suggestion?*

*Discovered a great weather gadget
you'd like to share?*

*Contact the Author,
Alan Fields
in one of four ways:*

1. By phone:
(303) 442-8792.

2. By mail:
1223 Peakview Circle,
Suite 800,
Boulder, CO 80302.

3. By fax:
(303) 442-3744.

4. By electronic mail.
My Compuserve address is
70312,124.
My American On-Line id is
ADFields.